LAKESHORE LIVING

LAKESHORE LIVING

Designing Lake Places and Communities in the Footprints of Environmental Writers

Paul J. Radomski and
Kristof Van Assche

Michigan State University Press

East Lansing

⊗ The paper used in this publication meets the minimum requirements of ANSI/NISO Z39.48-1992 (R 1997) (Permanence of Paper).

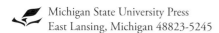 Michigan State University Press
East Lansing, Michigan 48823-5245

Printed and bound in the United States of America.

20 19 18 17 16 15 14   1 2 3 4 5 6 7 8 9 10

LIBRARY OF CONGRESS CATALOGING-IN-PUBLICATION DATA

Radomski, Paul.
Lakeshore living: designing lake places and communities in the footprints of environmental writers / Paul J. Radomski and Kristof Van Assche.
     pages cm
   Includes bibliographical references and index.
   ISBN 978-1-60917-408-8 (ebook)—ISBN 978-1-61186-118-1 (pbk. : alk. paper) 1. Lake ecology.
2. Environmental literature. I. Assche, Kristof Van. II. Title.
   QH541.5.L3R33 2014
   577.63—dc23
                                                                   2013023717

Book design by Scribe Inc. (www.scribenet.com)
Cover design by TG Design
Cover image of a great blue heron (*Ardea herodias*) taking a break on the dock after fishing in emergent vegetation along shore is © Andrea Lee Lambrecht, and used with permission. All rights reserved.

green press INITIATIVE Michigan State University Press is a member of the Green Press Initiative and is committed to developing and encouraging ecologically responsible publishing practices. For more information about the Green Press Initiative and the use of recycled paper in book publishing, please visit www.greenpressinitiative.org.

Visit Michigan State University Press at www.msupress.org

Contents

Foreword

RANDALL ARENDT

ONCE IN A GREAT WHILE A BOOK IS WRITTEN THAT CAPTURES THE ESSENCE OF ITS subject in a captivating and informative manner, and this is one of those rare volumes.

Readers will not only learn much from the well-researched text, but will enjoy the prose, which is fluid and retains one's interest, page after page. To discuss complex topics, and complex contributors to the field over the past seventy-five years, and render them readily understandable is a gift, not generally seen in books in the scientific or planning literature.

One of the aspects that makes *Lakeshore Living* such a pleasure to read is the way that the fascinating stories of Aldo Leopold, Sigurd Olson, and William Whyte have been related and intertwined with the technical themes eloquently discussed in detail in subsequent chapters.

If this were not enough, readers will also find practical solutions to many problems that have vexed and challenged lakeshore communities for decades, with pointers to specific techniques that have proven to be effective in other places where vision has combined with political will to overcome the unacceptable and unsustainable status quo.

As the principal purpose of a foreword is to describe what follows and to whet one's appetite, brevity is a virtue, and I do not desire to detain the reader from delving into this exceptional book, for which your hours of perusal will be well rewarded.

Preface

WE BELIEVE IN CHALLENGING THE STATUS QUO ON LAKESHORE DEVELOPMENT. WE believe in the need to rethink how we develop our lake lots, lay out our neighborhoods, and form our communities. We want to help people make beautiful places to live. This is a book for people who live on the shores or recreate on lakes, and it is for those with an interest in improving lakeshore living in a more sustainable and just manner. This book offers ways to live on the lakeshore and close to water and wetlands. It offers ideas on the creation of enduring lakeshore communities and on the principles of planning and design of lakeshore properties and neighborhoods—framed by the lives and thoughts of great environmental writers.

Our challenge is to learn, and to teach our neighbors, how to handle nature gently, especially at the land–water interface. This requires theory from the fields of ecology, limnology, landscape design, landscape architecture, urban planning, economics, sociology, and philosophy. Academics package this approach into systems thinking, systems theory, or systems science; here in the context of living well along the shore of a lake we just refer to it as rich lakeshore living. We need to think of lakes and lakeshores as systems, both ecosystems and social systems. The lake is part of the landscape. It is part of our economy, and it defines our web of relationships and how we live. For a lake home property to have a positive influence on the lake and for us to be successful in realizing a full, rich lakeshore life, we need to fight the desire to seek a single-solution approach or fall for the charms of all-explaining ideology or narrative. In other words, we'll have to distance ourselves in order to reflect and rethink. We believe this book can be part of such an endeavor.

It is easy to blame someone else for the problems associated with our lakes (e.g., farmers, loggers, and nonlakeshore residents) or to attribute our reduced lake quality to something outside us (e.g., the establishment of an invasive species and state or local agency failings). We don't like to hear that our behavior could also be a part of the problem. We must prepare. Growth of knowledge, as well as practicalities such as increasing energy costs, will force us to change, and we need to make the right changes to advance our quality of life. Societal events unfold slowly until a tipping point is reached. While there is still resilience in the system, we must reinvent our lakeshore living. Sometimes we need to go back into the past to retrieve the wisdom lost in the ever-increasing mass of information.

We can create a more enduring form of lakeshore living using a mixture of good ideas on community development from the past and techniques for sustainability from today's expanding fields of science. Factual information rather than ideology guides the pragmatist. We should discard our ideologies and unsupported beliefs. Ideology corrupts the mind and ruins communities. We should not confuse scientific skepticism with ignorance. Often we claim the first, when in truth it may be the latter. We need to be careful because the powerful exploit our ignorance. Scientific skepticism means doubting the claims that are made on something while being informed on the related facts. The open mind conducts an impartial search of the truth. The closed mind uses a partial search of the facts to defend a preconceived position. When to apply science-derived evidence is often a central dispute within communities. Decisions can be belief-based or politically-based when they benefit a class or group of people. However, shouldn't decisions be science-based when the consequences are critically important to everyone?

Influencing complex systems, like lake communities, requires first the understanding that the lake and its community are part of a larger system. It requires understanding the parts of the system, the strength of the connections between parts, the functions of the parts, the behavior of the integrated parts, the feedback loops, and the potential leverage points of the system. For this reason we describe lake ecology, human impacts on lake systems, and traditions of regulation. We propose design solutions for lakeshore developments that are implemented within the context of the natural landscape and the built world. Our neighborhoods will always be living systems built from a mix of human materials, nonhuman abiotic elements, and living organisms. This mix of built and natural elements is our habitat. Technology is important in creating this habitat, but building the connection back to ecology and the biological will improve our lives. We need to build nature in and around us, not with the technologies of nature simulation or imitation, but with the real stuff. All lake communities are inseparable from nature.

Ecologists, as well, will need to play a greater role in building nature in and around us. They may have to address their biases related to the inclusion of nonnative species, altering ecosystems, and study of city ecosystems. The landscapes between wilderness and city are sometimes denoted as "middle landscapes," which include farms, pastures, suburbia, and our lake communities. We do not see such a classification. We see a spectrum of wildness. On the domesticated end of the spectrum wildness must be added to improve the quality of our lives. To produce better places through the inclusion of nature, greater collaboration between ecologists, engineers, landscape architects, urban designers, financiers, sociologists, politicians, and citizens will be necessary. Whether designing a lake home place or a lake community, we've found that building effective cross-disciplinary teams is the key to better place designs.

Sustainable designs can only be site specific, incorporating a thorough site analysis and leading to a design that is sensitive to rare natural features, geology, hydrology, culture, people living in the neighborhood, and the architectural history of the region. We write from our own intensive experiences studying America's Midwest and southern Canadian lakes, dealing with shoreland development, and teaching about landscape architecture and European environmental planning; however, the facts and theories herein are broadly applicable to North American lakes and how we live along the shore. We must accept that changing just one aspect of the way we develop or live on the lakeshore will likely be insufficient to create a sustainable and just lakeshore-living model. However, the use of multiple tools—those that have been proven effective—and critical leverage points of our social, economic, and government systems might be sufficient.

This book was written in the belief that the environmental writing tradition, exemplified by the works of Aldo Leopold, Sigurd Olson, and William Whyte, can teach us something about rich lakeshore living. Their ideas and writings continue to capture people's imagination and provide insight and guidance on the fundamentals of rich lakeshore living. They grasped the fundamental issues of living close to the water, the multiplicity of linkages between ecology and society, and the paths toward better use. Each was a scientist, activist, and writer. For them systems thinking came after long study and numerous experiences trying to change systems. Along the way, their exploration for solutions was not limited by academic disciplines. We reinterpret their insights in the light of contemporary questions and combine them with recent scientific thought in ecology, planning, and related fields into a perspective about lakeshore living that can hopefully inspire scientists, politicians, residents, and lovers of literature and the environment.

Acknowledgments

PAUL RADOMSKI WOULD LIKE TO THANK HIS WIFE, HOLLIE RADOMSKI, FOR HER SUPPORT and encouragement while he spent evenings and weekends working on this book instead of out walking in the woods and enjoying her company. He would also like to thank Dave Wright and Dennis Schupp for their understanding and guidance, and Drs. Stan Szczytko, Irv Korth, and J. Baird Callicott for their teachings. Paul graciously thanks Kristof, who conceived the book and developed its initial foundation, for his support and friendship.

Kristof Van Assche would like to thank his parents for their patience.

We were very fortunate to have Kristin Carlson scrutinize the text, and we owe her a large debt of gratitude. We would like to thank the many people who shared their insights on lake and conservation issues with us along our journey of understanding, including, but not limited to Julie Aadland, Charles Anderson, Dennis Anderson, Adam Arvidson, Heather Baird, Larry Baker, Jim Ballenthin, Russ Barrett, Wayne Barstad, Romie Barwick, Ann Beaver, Tom Beaver, Jim Bence, Marian Bender, Angie Berg, Greg Berg, Lyn Bergquist, John Bilotta, Kristen Blann, Mary Blickenderfer, John Bogard, Gary Botzek, Joe Boyle, Tim Brastrup, Rick Bruesewitz, Jed Burkett, Andy Carlson, Phil Carlson, Ian Chisholm, Howard Christman, Yossi Cohen, Jean Coleman, Robb Collett, Steve Colvin, Carmen Converse, Sam Cook, Bill Darby, Harro de Jong, Dennis DeVries, Don Dewey, Randall Doneen, Gregg Downing, John Downing, Melissa Drake, Henry Drewes, Kate Drewry, Ryan Drum, Mike Duval, Harold Dziuk, Jeff Eibler, Bob Ekstrom, John Erickson, Linda Erickson-Eastwood, Joe Fellegy, Mike Findorff, Dave Friedl, David Fulton, Ed Fussy, Carol Gawlik, Dale Gawlik, Louie Gawlik, Ann Geisen, Jean Goad, Tim Goeman, Patty Gould-St. Aubin, Gerold Grant, Ken Grob, Jim Gustafson, Cindy Hagley, Nicole Hansel-Welch, Dr. Michael Hansen, Mark Hauck, Tom Heinrich, Steve Heiskary, Carrol Henderson, Don Hickman, John Hiebert, Steve Hirsch, John Hoenig, Mary Hoff, Pat Hogan, Dale Homuth, Mark Hove, Darrin Hoverson, Tom Hovey, Jeff Hrubes, Bill Huber, Phil Hunsicker, Dan Isermann, Pete Jacobson, Lucinda Johnson, Tom Jones, Larry Kallemeyn, Kendall Kamke, Anne Kapuscinski, Tim Kelly, Doug Kingsley, Beth Knudsen, Jim Kounkel, James Howard Kunstler, Jon Larsen, Kris Larson, Terry Lejcher, Jerry Lerom, Dave Leuthe, David Lick, Jim Lilenthal, Dale Lockwood, Dale Logsdon, Amy Loiselle, Dave Lucchesi, Chuck Marohn, Michael McDonough, Catherine McLynn, Jay Michels, Dave Milles, Jason Moeckel, Kent Montgomery, Gary Montz, Robert Morgan, Ron Morreim, Mike Mueller, Wayne Mueller, John Myers, Ransom Myers, Karen Myhre, Terry Neff, Darby Nelson, Tom Nelson, Ray Norrgard, Doug Norris, Jane Norris, Ben Oleson, Peder Otterson, Kevin Page, Bill Patnaude, Ron Payer, Mike Peloquin, Chris Pence, Don Pereira, Shawn Perich, Donna Perleberg, Ken Perry, Pam Perry, Dan Petrik, Lee Pfannmuller, Sharon Pfeifer, Ann Pierce, Rod Pierce, Phil Pister, John Postovit, Terry Quinn, Andy Radomski, Joan Radomski, Joseph Radomski, Noel Radomski, Rian Reed, Keith Reeves, John Ringle, Steve Roos, Chuck

Rose, Brian Ross, Jeff Schoenbauer, Kathy Schoenbauer, Jesse Schomberg, Russ Schultz, Johanna Schussler, Jennifer Shillcox, Molly Shodeen, Stephanie Simon, Terry Simon, Luke Skinner, Jack Skrypek, Doug Smith, Jeff Smyser, Byron Snowden, Paul Stegmeir, Brian Stenquist, Paul Stolen, Denise Stromme, John Sumption, Deb Swackhamer, Dan Swanson, Kimberly Thielen-Cremers, Lonnie Thomas, Bill Thorn, Dan Thul, Cindy Tomcko, Bob Valesano, Ray Valley, Henry Van Offelen, Bruce Vondracek, Jack Wallschlaeger, Judy Wallschlaeger, Chip Welling, John Wells, Reno Wells, Megan Wenker, Paula West, Julie Westerlund, David Willis, Bruce Wilson, Jack Wingate, Kevin Woizeschke, Allison Wolf, Jan Wolff, Rebecca Wooden, Tom Worth, Dave Zentner, and Molly Zins.

Quotations from *A Sand County Almanac* by Aldo Leopold (1966) came from Foreword, Marshland Elegy, Chihuahua and Sonora; Guacamaja, The Round River, The Community Concept, The Land Pyramid, The Outlook, and Wilderness for Science. They were used by permission of Oxford University Press, USA

Quotations from William H. Whyte came from the following books and are reprinted by permission of the Albert LaFarge Literary Agency, all rights reserved.

The Organization Man, by William H. Whyte. Originally published by Simon & Schuster in 1956. Copyright © William H. Whyte.

The Last Landscape, by William H. Whyte. Originally published by Doubleday in 1968. Copyright © William H. Whyte.

City: Rediscovering the Center, by William H. Whyte. Originally published by Doubleday in 1988. Copyright © William H. Whyte.

A Time of War: Remembering Guadalcanal, a Battle Without Maps, by William H. Whyte. Originally published by Fordham University Press in 2000. Copyright © William H. Whyte.

The Essential William H. Whyte, edited by Albert LaFarge. Originally published by Fordham University Press in 2000. Copyright © William H. Whyte.

LAKESHORE LIVING

Prologue

WE ARE NOT DEALING WITH A LUXURY PROBLEM AFFECTING ONLY THE LEISURE CLASS and lake lovers. We are dealing with a web of issues that we cannot afford to ignore. Failing to design our lakeshore places and to leverage the system changes that we will later identify will jeopardize existing lakeshore qualities and values, and we may miss the opportunities to create new qualities for lakes. Our lakeshore living needs reinvention. We believe in a better way of composing our lakeshore places. We believe in some basic, timeless principles that promote healthier shoreland places, friendlier neighborhoods, and sustainable communities.

How we live on the land is one of the most important environmental issues of our time. Where we live and how we are distributed defines sustainability, energy use, and quality of life.

Are you happy with the development around a lake you love? Are you happy with the quality of your neighborhood? What about your downtown? These questions often return answers indicating that things could be better. Lakes are being degraded, and we are depleting natural resources at a time when our ecological and social systems are less resilient. Increasing energy costs will create other challenges. Maintaining healthy lake ecosystems in the face of these changes demands a call to accelerate the advancement of design for lakeshore developments.

In this book, we present a renewed perspective on designing our lakeshore living, or, in fact, living near the water. This perspective is inspired by timeless principles. And these principles, we will show, were appreciated by generations of our most cherished environmental writers and conservationists. Why is living near the water so important? Because water itself is important. It is a resource for humans, it flows and connects everything in socio-ecological systems, and it structures an astonishing ecological diversity in networks of lakes, streams, wetlands, and invisible aquifers. Yet, before going to solutions, let's take a look at the main issues. What's going on?

POOR PLANNING AND NO DESIGNING

The planning regulations in North America are, on the one hand, a product of complex legal and political battles, changing with every new battle; on the other hand, they are traditional in scope, a set of minimum requirements designed to avoid substantial environmental damage. Designing places with sustainable new qualities cannot be deduced from these rules. Different analyses and different planning skills are needed. In the case of lakeshore living, many communities are planned and zoned in a very loose way, often with little enforcement, and with considerable tensions between the local rules and state/provincial rules. Even environmental laws and policies, which have a significant bearing on the possibilities for lakeshore redevelopment, cannot prescribe good design.

Before embarking on a lakeshore project, it is advisable to talk to local, county, and state/provincial officials to determine applicable rules. Those same conversations can help in understanding the ideology behind those rules and policies, as well as potential tensions and ability for negotiations. The political culture of the community will strongly influence the attitude toward planning in general, and particularly toward planning and environmental regulations. Still, in most communities, there is a strong awareness that things need to change, that the environment is declining, and that new development somehow needs to take into account environmental quality. On the developer side, there is also a growing awareness that the traditional "piano-key," where a series of long narrow lots ring the lake, or "lot-and-block" shoreland developments, with a maximum amount of parcels directly on the lake, are not the best use of space and not the most profitable. Developers and local officials are also aware that designs need to comply with the goals set at the regional level and that bending some of the newer and stricter environmental regulations will only be accepted when the design they propose does embody some of those goals. Planning and design is a negotiation, a give and take.

The main problem with the governance of shoreland development is that the outcome of a series of developments rarely meets the community's expectation. Paradoxically, development proceeds in a fashion inconsistent with local comprehensive plans. Vibrant development appears constrained by developments for automobiles (car habitat) rather than humans (people habitat). Conventional lot-and-block developments, or conventional subdivisions, are not preserving the lakeshore assets. Shoreland is often fragmented, with homes and docks every 100, 150, or 200 feet regardless of vulnerable or unique natural features or conditions. Conventional subdivisions essentially produce only lots and roads. They provide few places to interact with nature and neighbors. This development approach comes with a high cost of community services. According to the American Farmland Trust, for every dollar of tax revenue raised from such residential developments, it costs on average $1.16 in public services. Reliance on conventional subdivisions creates community deficiencies—no community social places, no neighborhood amenities, and no walking and biking trails. Shoreland planned-unit developments were envisioned to achieve great benefits with clustering of homes and protection of open spaces; however, most local ordinances have 1970s-era open-space standards, which are often ambiguous and weak. Thus, many of the open-space amenities were never realized.

An intelligent and far-sighted land development and redevelopment ordinance is an essential element in good land-use planning. Standards should reflect, complement, and support the community vision for itself and its comprehensive land-use plan. However, our problems with the governance of shoreland development go beyond failed planning and misguided zoning. Most importantly there is no or little shoreland community design. The look and function of many of our lakeshore communities give the impression that we do not care for beauty, people, and nature.

DIMINISHED WATER QUALITY AND ECOSYSTEMS

There is widespread concern about the consequences of poor development on water quality and fish and wildlife habitat. Lake development pressure is increasing with more dwellings per lake each year. More people are choosing to live and recreate in the lake-rich districts. These areas are likely to see a large influx of people. Population increases, along with the associated loss of uninhabited shoreland areas, has led to a greater public concern. Limnological data support many of these concerns and perceptions. As a lake's watershed (lakeshed or area of land that drains into a lake) becomes more cultivated or urbanized, nutrient levels increase and water clarity decreases due to pollutant runoff, poor stormwater management, and shoreline phosphorus inputs from shoreland sewer systems and lawns to the lake. Human habitation along the shore usually has a cumulative negative effect on fish and wildlife habitat, water quality, and biota of lake ecosystems. Initially the greatest impact of this shoreland development is habitat alteration, which results in the decline of fish and wildlife populations.

Agriculture's Unintended Consequences

Long ago our ancestors randomly began tinkering with gardening and with the tending of the tamest of indigenous herbivores. Those groups that tinkered and tended successfully began to prosper, and for them a dramatic lifestyle change followed with specialization of individual roles within larger communities. This way of life has become the dominant culture on our planet rather than the countless other ways of living (some of which are still practiced by holdout descendents). The consequences of this agricultural lifestyle epidemic can be understated.

We consume a large percentage of the Earth's net amount of biomass produced each year by plants. Years ago, scientists estimated that we appropriated 25 to 40 percent of the total primary production of the planet's biosphere, and in some regions up to 80 percent. This estimate caught the attention of people because if we don't learn to share or set aside a large portion of the planet's productivity for others, then the fate of many wild animals is sealed and, more importantly, we ourselves will soon be in big trouble. We take an astonishing share of the world's productivity. Our gardens and grazing lands now occupy more than 35 percent of the Earth's arable land, and these domesticated lands are now the most common terrestrial ecosystem type on the planet.

Agricultural lands are increasingly highly engineered ecosystems. They are systems with extremely low biodiversity, drained with ditches and tiling—their high productivity dependent on both good soils and large inputs of nutrients and pesticides. These lands shed large amounts of sediment and nutrients to the air, neighboring lands, and adjacent waters (of course not all agricultural lands contribute equally to these pollution problems). In the last seven water-quality reports to the U.S. Congress, the U.S. Environmental Protection Agency has identified agriculture as the top or one of the leading causes of water-quality impairment of rivers and lakes in the United States.

Nutrients reaching the lake result in eutrophication or an accelerated rate of lake aging. Eutrophication conditions include higher occurrence of noxious algae blooms, excessive plant growth, loss of water clarity, and low dissolved oxygen. Small additions of phosphorus, a plant nutrient that is rich in many North American soils, can lead to large reductions in clarity. Many lakes will fail to recover even after excessive nutrient additions are eliminated. Eutrophication can alter nutrient cycling dynamics within a lake when large volumes of anoxic, or oxygenless, water in the hypolimnion (i.e., the bottom water layer in a lake) are created by bacterial consumption of dissolved oxygen.

Failure to Manage Rainwater

Lakeshore development also increases nutrient inputs to lakes. Shoreline development (impervious surfaces and lawns) increases both the amount of runoff and the quantity of pollutants and nutrients reaching lakes and rivers. Shoreland development and drainage basin alterations have resulted in long-term declines in lake water quality. Because sediment naturally builds up on a lake bottom over time, an accurate record of environmental change can be found in the lake's sediment layers. Lakes store information. Bruce Wilson, a Minnesota lake expert, stated "lakes have a long memory—it's called sediment." Analogous to interpretation of the growth rings of a tree where climatologists and foresters can learn about past weather and forest change, scientists can reconstruct the past conditions of a lake from its accumulated sediment. Paleontologists drive plastic tubes into the bottom sediments and bring up a core to be analyzed. Researchers have found that tiny algae called diatoms live under very narrow environmental conditions. If the water quality is poor, not all species of diatoms can survive, so the species of diatoms present in the sediment are good indicators of past water quality. Using these paleolimnology techniques scientists have documented the consequences of shoreland development on lake water quality. These studies usually show several key events that seem to be shared by many lakes.

- In many lakes, there was an increase in lake sediment accumulation in the early twentieth century due to logging and other land disturbances.
- The initial shoreland development on a lake generally had minimal impact on lake water quality.
- The highest sediment accumulation often occurred during the peak construction phase of converting shoreland cabins to year-round homes. In many low-alkalinity lakes, water clarity decreased with development.
- In undeveloped or lightly developed northern lakes, phosphorus levels and water clarity remained unchanged from 1750 to the present. In developed lakes, however, substantial increases in phosphorus levels and resulting decreases in water clarity were found due to urbanization or agricultural activities in their watersheds.

Erosion and sedimentation are the obvious effects of failure to manage rainwater, which in turn trigger a series of processes that reduce water and habitat quality. Stormwater runoff is considered a major source of water pollution and may be responsible for up to 15 percent of river and lake water impairment in the United States. Perhaps the single greatest threat to lakes from sediment is as a carrier of phosphorus to the lake. Sediment delivery from agricultural areas, timber harvesting areas, and construction sites can be a major source of pollution. Land disturbing activities, like grading or filling of even small amounts of material, generally have serious potential to harm the shoreland. Shoreline vegetation is usually not adequate to prevent this pollution from reaching the lake; however, proper erosion and sediment control practices in agriculture, forestry, and construction zones can reduce this problem.

In residential areas, rainwater runoff originates from streets, driveways, parking lots, roofs, and other impervious surfaces. Rainwater that does not infiltrate into the ground or evaporate runs downhill to lakes, wetlands, or rivers. As impervious surface coverage increases, the amount of nutrients entering waters increases linearly. When impervious surface coverage exceeds 10–12 percent, without a comprehensive approach to manage rainwater, water quality is often negatively impacted. There is also a definitive link between impervious surface cover and fish assemblages. Sedimentation and toxic pollutant runoff to streams and lakes increase with imperviousness and lead to reduced fish reproductive success and survival. In northern communities, the winter use of salt for road deicing results in increasing sodium and chloride concentrations in lakes, which at high concentrations can harm plants, frogs, fish, and other organisms.

In residential areas, the largest source of phosphorus entering lakes comes from lawn and impervious surface runoff. Rainwater runoff from developed lawn-to-lake managed shoreline was measured five to ten times higher than from forested shorelines. Runoff from lawns occurs more frequently than previously thought, with a high percentage of storms resulting in runoff. Lawns and urban soils are often very compacted and may act like impervious surfaces in increasing rainwater runoff. Many lakeshore sites with lawns to lake have been heavily graded during construction. The depressions and swales that would normally retard runoff are often graded over, the topsoil removed, and the underlying soil compacted, making a flat lawn. But flat lawns are more like pavement in their inability to infiltrate and retard stormwater runoff. Water flowing over lawn surfaces picks up dirt, nutrients, pesticides, toxic chemicals, pet waste, and other pollutants. In addition, the lawn-to-lake shoreline allows seven to nine times more phosphorus to enter the lake than an undomesticated (naturally) vegetated shoreline. While absolute values of phosphorus entering the lake from a developed shoreline lot vary due to soil, slope, and other site-specific conditions, a lawn-to-lake lot has been estimated to average 0.2 pounds of phosphorus per summer compared to 0.03 pounds per summer for a lot with a naturally vegetated shoreline buffer. Groundwater under lawn areas can also have high concentrations of nutrients. Hydrologists have found nitrate and total phosphorus concentrations three to four times higher in groundwater under lawn areas compared to wooded areas. Sediment from heavy runoff over lawns and the erosion of shoreline can itself be a pollutant, because it can impair the feeding and reproduction of many forms of aquatic life. The sum of individual lot contributions to total runoff pollution accumulates around a lake, often creating serious lake water quality problems.

Loss of Shoreline Buffers and Habitat

Shoreline buffers are corridors of diverse vegetation along rivers, streams, and lakes that help protect water quality by providing a transition between upland development and adjoining water. Abundant, diverse vegetation traps, filters, and impedes runoff. Shoreline buffers stabilize lake and river banks, offer scenic screening of shoreline development, reduce erosion, control sedimentation, and provide habitat for shoreline species. The mismanagement of this shoreline buffer adversely impacts the natural resources of shoreland areas.

Trees, shrubs, and the forest understory near the shore have declined over time along some developed shoreline. This change in shoreline habitat affects the structure of bird communities. Common suburban-style birds like chickadees, cowbirds, blue jays, and grackles replace the uncommon "species of special concern" birds like loons warblers, and vireos. The probability of loons on a lake decreases with increased riparian housing density. Loons will not likely nest on a groomed and manicured beach—they prefer to nest near shore on vegetated hummocks, small islands, or masses of emergent vegetation. The loss of trees along shore means fewer trees that fall into the water. Fallen trees provide habitat for fish. Biologists have determined that this loss of trees due to development will negatively affect fish populations

©ANDREA LEE LAMBRECHT

A COMMON LOON NESTING ALONGSHORE.

for centuries. Downed trees provide important in-lake structure, habitat, food, and shelter for fishes, frogs, turtles, waterbirds, and mammals. This woody habitat is also important for aquatic invertebrates like snails and dragonflies. Nearshore downed trees also blunt waves and ice action that scours the shoreline.

Green frogs, common along shorelines in the past, have disappeared in areas with high development. Male green frogs establish breeding territories within several feet of the lake's edge, but disturbance to the shoreline vegetation eliminates their habitat. It is these critical nearshore areas that are often altered or destroyed. While many informed lakeshore owners leave or restore native vegetation along the shore, a large number of homeowners mow a lawn down to the lake. This lawn-to-lake management style fragments the nearshore habitat. Fragmented habitat forces frogs and other wildlife to spend extra time and energy seeking access to nesting, basking, and feeding sites. Changes in aquatic plant communities may also occur with lakeshore development. Poor treatment of the shoreline buffer often corresponds to poor treatment of the aquatic habitat. Developed shorelines often have less floating-leaf and emergent vegetative cover than undeveloped shorelines.

Inadequate Lakeshore Sewage Treatment

For many lakes, wastewater from lakeshore developments goes into individual subsurface sewage treatment systems, commonly called septic or sewage systems. Such systems consist of a tank and drain field (i.e., an absorption area). The tank captures solid material and anaerobic bacteria decompose some of the solids. The wastewater that leaves the tank, or effluent, contains significant amounts of pathogens, pollutants, and nutrients. The drain field, with a system of perforated pipes, distributes the effluent to a large area so that aerobic bacteria can further break down pathogens and the soil can absorb phosphorus and filter the effluent.

On-site sewer systems are prone to failure. Phosphorus pollution from sewage systems is the main concern. One pound of phosphorus can produce five hundred pounds of algae. A household produces about two pounds of phosphorus per person each year, and in rural areas it is commonly discharged into a subsurface sewage treatment system. Conventional subsurface sewage treatment systems can be effective at removing phosphorus. Drain-field soils usually absorb or mineralize phosphorus; however, certain soil conditions and close proximity of drain fields to lakes can result in phosphorus pollution. In addition, the capacity of the soil to retain phosphorus is finite, and phosphorus movement deeper into the soil profile and closer to water resources can be expected. Researchers have found that elevated phosphorus concentrations in groundwater are usually within fifty feet of functioning sewage systems. However, some phosphorus plumes have been found to extend beyond sixty-six feet from drain fields. Other evidence suggests that drain fields should be at least one hundred feet from the lake to minimize the risk of phosphorus reaching the lake; for sensitive lakes with lakeshore soils that are conducive to nutrient movement, this distance should be much greater.

Unlike with phosphorus, conventional sewage systems are relatively ineffective at removing nitrogen. Nitrogen (in the form of nitrate) dissolves easily in water and is stable over a wide range of environmental conditions. It is highly mobile and easily transported in groundwater, flowing through the soil and often ending up in lakes or even well water. In the city of Baxter, Minnesota, a study found nitrate levels down-gradient of sewage drain fields exceeded the drinking-water criteria at all sites surveyed. In addition, nitrate concentrations in domestic wells from three central Minnesota communities increased with increasing age of the sewer systems in the area. Sewage systems may contribute significant nitrogen to lakes even when in compliance with the regulations because they are rarely designed to remove nitrogen. Concentrations of nitrogen in sewage tank effluent are generally two to ten times higher than the common drinking-water standard for nitrate-N (e.g., ten milligrams of nitrate-N per liter). Unless drain fields are much larger than required by code or alternative treatment systems designed for nitrogen removal (e.g., constructed wetland or certain recirculating filters) are used, less than 25 percent of the nitrogen will be immobilized in the soil. The remainder ends up in the lake via direct drainage from subsurface seepage and by intermittent flushing of surface seepage by rainstorms or snowmelt runoff. Nitrates and phosphorus that get into the lake may lead to increased aquatic plant and algae growth, including algal blooms, decreased oxygen levels, and eutrophication.

A PATH TO LAKESHORE DESIGNS

The list of problems is longer. Even after this brief introduction, it may be clear that the issues confronting our lakes are complex and convoluted, and that people living near the water are not simplifying the matter. Yet people are part of nature, and for people, lakes are important.

We believe that our perspective can help communities and individuals to manage their places, lakes, and territories in a more enduring way.

Development cannot and should not be stopped around every lake and wetland. We need better ways to accommodate both people and the environment in these fragile areas. Design is part of the solution, because in design we can coordinate decisions, people, and resources, and we can preserve and create spatial qualities. We can call this approach ecological design or green urbanism, or just lakeshore design—context sensitive, site specific, and multifunctional.

We will outline principles of lakeshore living expressed by Aldo Leopold, Sigurd Olson, and William Whyte. We will blend those principles with sustainability principles derived from our own unique experiences and learning—one from lake ecology and shoreland development governance and the other from landscape architecture and European environmental planning. When we speak of sustainability, we refer not only to a vision, but to a goal. Sustainability is the conservation of natural resources, and it is the striving for a better relationship with our world.

We unfold our perspective in a few steps. First, we ask ourselves the question what a lake is, a useful question if we want to address some of the issues mentioned above. We present the main elements and creatures of lakes and the surrounding landscapes and unravel the most important relations that turn these landscapes into ecosystems. Then we review the lives of Leopold, Olson, and Whyte. Their scientific and cultural appreciation of lakes and lakeshore living lays the groundwork for the next chapters, where we articulate a strategy toward sustainable lakeshore living that sees development crucially as redevelopment sensitive of ecological, cultural, and policy contexts. Asset mapping, assisted by the scientific insights from the first chapters, leads into asset creation and further to suggestions for connecting assets, old and new, into lakeshore communities for humans and the rest of nature. Finally, we dwell on the importance of existing policies, where we hark back to insights from the writers and in the cultural embeddings of lakeshore living.

At the end, we articulate reasons why local governments should do less traditional planning and zoning. Planning implies that local governments are controlling development, but this is rarely the case. Zoning implies dividing a county, town, or city into designated uses and applying development ordinances to each use area; however in many jurisdictions, zoning may have gone too far. We advocate two critical points. First, that we design our lakeshore homes and landscapes consistent with the articulated strategies of sustainable lakeshore living. Second, that local governments facilitate environmentally sensitive, human-scaled community-designing processes and that they resume the responsibility for large-scale design that focuses on designating the layout of all streets and roads.

The reader will find in the extensive notes and literature sections a wealth of historical information on secondary characters and events in our narrative of lakes, writers, and lakeshore living, as well as more detailed scientific data and policy concepts underpinning our reasoning and recommendations. For those inspired to plunge into lakes research or, even better, to use this book in design, policy making, or community discussions on lakeshore living, we recommend taking a look at these sections.

PART ONE

North American Lakes

WHILE A RIVER APPEARS TO CHANGE WITH ITS FLOW, WE GENERALLY PERCEIVE THAT a lake only changes with the seasons. We often interact with the lake in the summer, so we may have some understanding and appreciation for summer lake ecology. However, winter lake phenomena are often hidden from us, with the comings and goings of lake plants and animals unknown and underappreciated. Understanding our lake increases our sense of place and our affection for lakeshore living.

With colder weather the lake and shoreline environments change. Water, which has an unusual molecular structure, becomes lighter as it cools below 4 Celsius (39 Fahrenheit). Ice floats rather than sinks; can you imagine what Earth would look like if ice settled to the bottom of the lake instead of the surface? As ice forms, the lake is sealed off, eliminating substantial air exchange and the forces responsible for water circulation. The ice sheet itself is dynamic. The thundering sounds of lake ice on a cold night are eerie, with pops, booms, and loud crackings. The ice sheet expands as it warms, resulting in strong forces that push ice up on shore or create a line of cracked ice across the middle of the lake.

The coldest water is just under the ice, and the warmer, denser, 4C water is at the lake bottom. Snow on the ice dramatically reduces the amount of light that reaches lake plants, making the lake a dark, cold place. In shallow, fertile lakes this reduction in light can be fatal to fish that depend on oxygen dissolved in the water. In many lakes there are areas where dissolved oxygen disappears during summer and winter. These areas are in the fertile basins, near the bottom, where aerobic bacteria of decay consume available oxygen. It does not pay to fish in these basins; the fish have moved to oxygen-rich areas.

Aquatic plants keep a lake alive. Much of a lake's dissolved oxygen originates from algal and vascular plant photosynthesis. In addition, plants provide cover and food for the animals. Thus the fate of our lake plants is critical for our lakes. Many pondweeds, which are perennials, die back in the fall. Water lilies also die back, sprouting again in the spring from their roots. However, some pondweeds and other aquatic plants act as winter annuals or continue to grow throughout the winter. Coontail (*Ceratophyllum demersum;* a perennial, submersed plant with finely dissected whorled leaves and no roots), Canadian waterweed (*Elodea canadensis;* a perennial, submersed herb with leaves in whorls of three), largeleaf pondweed (*Potamogeton amplifolius;* a perennial, submersed herb with large, arced leaves), and Robbins' pondweed (*P. robbinsii;* a perennial herb with stiff leaves arranged in a fern-like appearance) can often be found growing and doing quite well under the ice. Algae do not grow much during the winter, but spring and fall diatom blooms can be spectacular. Diatoms are beautiful plants that live in silica houses; you must see them under a microscope to believe in their existence.

The aquatic insects have also adapted to dark, cold lakes. Dragonfly nymphs are hanging on to remaining plants, feeding little during this time. The whirligig beetles (Gyrinidae), insects that glide on the water surface, sometimes in large swarms, have buried themselves in the lake mud to hibernate. They will come out in the spring and lay their eggs on the pondweeds. The midges, nonbiting mosquito-like flies that swarm around the cabin and die en masse on the spiderwebs under the eaves, have laid their eggs in the lake this fall; the next generation will grow up when the water begins to warm, and we'll sweep their dead bodies off the windows again next year.

Fish behavior in winter has been studied for hundreds of years by interested anglers and biologists. As winter progresses, fish move to waters where there is oxygen. Some fish can be semidormant during this season. Others, like walleye (*Sander vitreus*), are active and feed throughout the winter. Crayfish move to deeper water in the winter, but the frogs, turtles, and snails are buried in the mud hibernating. Other animals, like bryozoans, live more bizarre lives. These tiny invertebrates, which prefer to live in clean water, form large colonies consisting of thousands of individuals. The colonies are gelatinous and spherical in shape, and they can reach the size of a beach ball. Bryozoan colonies die in the fall, but before they die they produce statoblasts. A statoblast is a tough, small, overwintering egg-like life stage. Bryozoan statoblasts germinate in the spring and produce new colonies that can be seen in nearshore areas attached to fallen tree branches or other structures.

The shoreline changes with the seasons as well. In the fall, trees along the shore drop their leaves with some entering the lake to feed leaf-eating aquatic invertebrates and to enrich the lives of leaf-decomposing fungi and bacteria. The memories of family fun and the liveliness of bald eagles (*Haliaeetus leucocephalus*) at their lake home nest are fresh, but our memories of the sounds of spring with the calls of killdeer (*Charadrius vociferus*) and other birds have faded. Of the seasons on the shore winter is quietest, but natural shorelines have abundant cover for overwintering animals. Chickadees (*Poecile*) and nuthatches (*Sitta*) can be seen in the winter searching for food on the trees and bushes. Rabbits, hares, squirrels, mice, and deer browse on last year's vegetative growth or dig in the snow for acorns. Pileated woodpeckers (*Dryocopus pileatus*) can be observed removing chunks of old dead big-tooth aspen trees (*Populus grandidentata*) as they search for ants and beetle larvae. Though merely shells of their former selves, these dead trees continue to stand at the shoreline until decay and gravity topple them into the lake, where they provide shelter for fish, turtles, and ducks. Bulrush (*Schoenoplectus*) stems, brown with roots dormant in winter, protrude out of the ice and snow and outline where the land ends and the water begins.

Knowledge of the environment surrounding us can bring us closer to a place, can render us placed people. Knowledge is also critical if we want to live in a way that accommodates both people with widely diverging interests and the rest of nature. In the case of lakes, this knowledge includes a thorough understanding of watersheds and the circulation of water, nutrients, and energy through these systems. In later chapters, we will show that this includes knowledge regarding sustainable planning and design, planning and design capable of maintaining the integrity of these ecosystems. First, however, we need to introduce important lake components and lake ecology. Just as Leopold, Olson, and Whyte, we believe an understanding of lakes, landscapes, and people inhabiting them can help us design with ecosystems in mind.

CHAPTER 1

Lake Parts

IT HAS BEEN ESTIMATED THAT THERE ARE ABOUT THREE MILLION LAKES GREATER than 25 acres (0.1 square kilometers) on the planet. These lakes are not distributed evenly over the world's landmasses. Earth's north temperate zone, including North America, is lake rich; Minnesota is called the land of ten thousand lakes, and Finland is called the land of thousands of lakes. Canada has over thirty-one thousand large lakes, and about 9 percent of the country is covered by freshwater. In North America, the highest densities of lakes are in the northeast and areas associated with glaciation. The continent's lakes are diverse in both size and character, ranging from small, fertile water bodies to the Great Lakes. People through the ages have been attracted to lakeshore living. We are attracted to lakes for food, home, and solace.

PHYSICAL FEATURES OF LAKES

Lakes are places. Lakes are made up of living and nonliving things. They are more than pools of water. Lakes are ecosystems that connect to other systems. Understanding lakes begins with understanding their basic elements. First, a lake includes a large amount of water, though what constitutes a large amount of water is arbitrary. The line between lakes and ponds is a fuzzy one. If you've spent time on Lake Superior, the world's largest lake by surface area, your perspective may be different than someone who has never experienced big water. Ponds are often characterized by shallow water, where light penetrates to the bottom, and they lack waves. Using this definition of a pond, Henry David Thoreau's Walden Pond, located in Concord, Massachusetts, is actually a lake. (Every New Englander might know that Walden Pond is a lake, but its name can confuse people from other areas.) Walden Pond is a deep (102 foot; 31 meter), small (61 acres; 25 hectares) lake formed by glaciers over ten thousand years ago. Most of the world's lakes are small; however, about 9 percent of the lakes account for about 60 percent of the total lake surface area.

Second, a lake's boundary is defined by its shoreline. Shorelines are dynamic places. Waves smooth out irregularities and deposit fine sediment in quiet areas, and water levels rise and fall with changes in the hydrologic cycle. At the open shore, trees and shrubs fight to reach the sun and to hold the soil at their roots. Lakes in a glacial outwash plain may have sand and soft sediment shorelines, whereas lakes situated in areas of glacial till often have rocky shorelines. The length of a lake's shoreline is dependent on the scale of measurement (measuring with a ruler produces an estimate larger than that measured with a yardstick), and based on the mathematics of Benoit Mandlebrot, one could also say that the length of a lake's shoreline is nearly infinite. With regard to fish and wildlife habitat, the shoreline generally refers to the narrow band around the lake centered on the land–water interface.

Third, subsurface, surface, and atmospheric water systems contribute water to a lake and can be considered parts of a lake. Beginning with these basic elements, lakes have a rich set of physical features that we can explore. Lakes receive water from precipitation, inlets, and groundwater, and lose water by evaporation, transpiration, outlets, and seepage to groundwater. Hydrologists can determine a lake's water balance by estimating inflow and outflow from all sources. Except for very large lakes, precipitation contributes generally only a small proportion of the water received by a lake. Drainage lakes, as their name implies, receive and lose most of their water from surface water inflows and outflows (inlets and outlets). Groundwater provides an important source of water for lakes in glacial till areas. Such lakes extend below the groundwater level, and the water seeps into the lake by percolating through the lake sediments or enters at discrete springs in the shallow water areas along the shore. Seepage lakes have no inlets and outlets and receive their water mostly from groundwater. Spring lakes are similar but have outlets that form the headwaters of a stream network.

In their studies of productivity, nutrient cycling, and animal and plant communities, limnologists often divide the lake into zones. This defining of lake zones is based on light transmission. The amount of energy received by the lake is dependent on the angle of the light as it hits the water. Latitude, season, time of day, and wave action all affect the angle. A substantial portion of the energy is often reflected. On a clear summer day, about 6 percent of the total radiant energy is reflected. The light that enters the water is either scattered or absorbed. The infrared portion of the energy spectrum is absorbed primarily by the first three feet of lake water, so only the top of the water column becomes sun-warmed. Turbid and dark-stained (from humic acids) water have higher absorption rates and thus lower transmission and transparency. Algae and other aquatic plants use the scattered, diffuse light to create their own food in the process of photosynthesis—a process that releases oxygen as a waste product. Aquatic plant production is reduced with increasing water turbidity, and algae production can limit water transparency. In 1865, Angelo Secchi, an Italian scientist, developed a simple index of a lake's transparency of light. He lowered a white disk into the lake and recorded the depth at which it was no longer visible. While photometers for measuring transmission of light through water are readily available today, the Secchi disk is still used. A common standardized Secchi disk method for lakes uses an eight-inch (twenty-centimeter) disk with black and white quadrants that is lowered from the shaded side of the boat during midday. The observer records the mean depth of the point at which the disk disappears during lowering and the point at which it reappears upon raising. Secchi disk observations range from a few inches in very turbid waters to over one hundred feet (thirty meters) in crystal clear waters.

The lake zones often used by scientists include the littoral, profundal, and pelagic zones. In the littoral zone, light is available for aquatic plant growth. The pelagic zone is the free open-water area of a lake, and the profundal zone is that part of the basin where light does not penetrate. Most of the lake productivity occurs in the littoral zone. The profundal zones are areas of decay and provide scientists with abundant information on lake chemical and historical characteristics. The pelagic zone is where offshore plankton, fast boats, and kids on tubes hang out.

The littoral zone is also highly influenced by wave action. Waves result from the force of wind and the friction of the water. As wind pushes the water, the water moves in the orbit of the wave but not in the direction of the wave. The maximum height of waves on a lake is a function of the fetch, or the uninterrupted distance across water in the direction of the down. As waves enter the littoral zone's shallow water, their speed decreases due to drag on the lake bottom. As the wave slows, the wave height increases and the wave becomes unstable, resulting in a wave break. The energy in the wave is now directed to the lake bottom and the shore. In areas with glacial till, the erosion at the foot of the shoreline bank brings additional rock down to shore. The process continues and the bank slowly moves landward, which creates a horizontal bed of rock alongshore. Natural beaches exist where sand is exposed by waves or where the wind and waves deposit the sand they have carried. Along high-energy shores, waves keep the beach clear of finer sediments and the sand in constant motion. These high-energy shorelines are often lacking in aquatic vegetation, as the continual wind and waves make it difficult for aquatic plants to colonize. Where gravel and rock are present in deepwater lakes, high-energy shorelines provide important spawning habitat for walleye, white sucker (*Catostomus commersoni*), cisco (*Coregonus artedi*), lake whitefish (*Coregonus clupeaformis*), lake trout (*Savelinus namaycush*), and burbot (*Lota lota*).

Lakes may also be divided into zones based on temperature and chemical composition. The sun warms the water close to the surface. Wind mixes this top water, creating a layer of uniformly warm water across the lake for a considerable depth. Below this is a pool of cooler water. This phenomenon of temperature variation with depth is called thermal stratification. The transition from warm to cool water, which is often abrupt, is known as a thermocline. Swim down into the depths of a deep lake in midsummer and you will soon realize when you have entered that cool, nonmixing pool of water. In deep lakes this thermal stratification lasts all summer. In shallow lakes the stratification may break down with high winds, only to return during calm periods (polymictic lakes). Many lakes circulate twice a year (dimictic lakes), in the spring and in the fall. A lake is said to turn over when the entire volume of lake water circulates.

The mixing of water, or lack thereof, can have a profound impact on dissolved oxygen concentrations in the water. This is important, as dissolved oxygen is vital to aquatic organisms from fish to bacteria. Oxygen diffuses in water from the air and from algae and aquatic plant photosynthesis. In lakes with very low nutrient concentrations and organic matter production, the oxygen concentration remains close to the saturation point at all water depths. In more productive lakes, the oxygen concentration progressively decreases in the deeper water after spring turnover. Bacteria in the bottom sediments consume the oxygen as they eat the accumulated organic matter on the lake bottom, converting much of the dead tissue into carbon dioxide gas and water vapor. In very productive, or eutrophic, lakes, dissolved oxygen can be depleted enough to create an anaerobic condition for much of the summer. This anaerobic condition allows the release of phosphorus, iron, and manganese from the lake sediments. Fish need high dissolved oxygen concentrations, so they spend most of their time living at depths where they find this. When oxygen is depleted at the lake bottom, aerobic bacteria are replaced with anaerobic bacteria and decomposition rates slow down. Anaerobic bacterial decomposition produces methane and carbon dioxide gases. In the summer, as methane diffuses up from the lake bottom into oxygenated waters, aerobic bacteria consume the methane and produce carbon dioxide gas as a waste product. Sometimes one can observe a mass release of methane and carbon dioxide gas: enter a muck-filled bay during an extended calm period, disturb the lake bottom, and witness the gas bubble to the surface.

Caused by wind or temperature gradients in the water, currents in the lake surface water are common. The phenomenon of Langmuir circulation can be observed on lakes when wind speeds are between five and sixteen miles per hour (two and seven meters per second). Langmuir circulation is characterized by parallel surface streaks of aggregated matter aligned with the direction of the wind. These windrows of seeds, leaves, and other plant material are the result of wind-produced surface currents that create a series of helical currents in the surface water. In the Great Lakes, large circulation patterns also produce slicks of aggregated material, but they are of greater size and separation than the streaks created by Langmuir circulation. Seiches are another fascinating phenomena of lakes. A steady wind causes water to pile up at the downwind end of the lake. This forces the tilting of the cooler, dense deep water deeper on the downwind end and shallower at the upwind end. The water level will be higher at the downwind end of the lake than at the upwind end. When wind stops, the warm, less dense surface water that piled up sinks until it hits the cooler, dense deep water, then flows horizontally back from where it came. In a thermally stratified lake, the thermocline may teeter back and forth for hours or days before returning to an equilibrium state or horizontal condition. Water moves within lakes even when they are covered with ice in the winter. These winter water currents are the result of heat rising from the lake sediments and from underwater springs.

The selection of plants and animals listed below is not meant to replace field guides but to provide a more detailed sketch of lake ecology. While we interpret the ecological value of particular species or taxa, we could remember that they are things to admire and appreciate.

PLANTS

A lake's plant community provides many environmental services to us, such as absorbing nutrients that reduce water quality, reducing erosion from waves, and providing food and habitat for fish and wildlife. Perhaps as important, the native flora, more than anything else, defines the ecological character of our lakes. The animals we see from our dock or cabin add to the experience of lakeshore living, as well as contributing to the ecological processes of the lake.

Blue-green cyanophyta (bacteria) are primitive plants that have existed on Earth for billions of years. They produce a considerable amount of oxygen through photosynthesis. Blue-greens come in many colors, from red to paint-like greens and violets, and they may exist as single cells, multicell filaments, or large colonies. In productive systems, blue-greens grow rapidly and extensively, creating noxious scum or slick conditions. They reach their highest abundance, often called "blooms," in mid to late summer. These blooms can produce strong odors, and they contain toxins that cause rashes, respiratory distress, and other problems. Wildlife and pets are also highly susceptible to these blooms; the cyanophyta-produced toxins can cause death with very high exposure. Common genera of blue-greens include *Microcystis* and the filamentous genera *Anabaena* and *Aphanizomenon*. Some organisms, such as certain zooplankton species, eat blue-greens; however, their contribution to the lake food web is generally as a source of food for bacteria after they die and settle on the lake bottom.

Algae, like the blue-green cyanophyta, exist as single- or multicelled life forms. Algae can be found in the open water, attached to aquatic plants, and even on the lake bottom if light is sufficient. Algae form the foundation of a lake's food web, as planktonic algae (phytoplankton) are the main food source for microcrustaceans and other small planktonic organisms (zooplankton). In the littoral zone, algae provide food for insects and other organisms, such as tadpoles, salamanders, and fish. Families of algae include green algae, golden-brown algae, and diatoms. There are thousands of species, and the variety of form is quite spectacular, ranging from the beautiful filaments of the green algae *Spirogyra* and the spherical colonies of *Volvox* in shallow water, to diatoms with silica cases and petri dish–like structures like the pinnate *Aneumastus* and the centric *Stephanodiscus* and *Cyclotella*.

A large number of algae species can be found in a lake at any given time, with some species reproducing and others in resting stages. Algal communities are very dynamic, fluctuating with light and temperature conditions, availability of nutrients, competition, and grazing by small organisms. If a lake is turbid or not very transparent, most of the algal activity will be near the surface. In other lakes, algae that are adapted to low light and cooler water temperatures may be most productive at greater depths. There is often a seasonal rhythm to algal activity. Winter is a period of low growth due to low light conditions and cold water. In spring, the large diatom algae dominate due to their preference for cool temperatures. Diatom blooms might last a few weeks to a month. Not surprisingly, the availability of silica in the lake water can influence the abundance of the diatom algae. During summer in productive lakes, various green algae proliferate followed by blooms of blue-greens. Diatoms may again become dominant in the fall, with their activity abruptly declining when winter comes.

The annual variability of algae productivity in a lake is often determined by nutrient loading. For most lakes, phosphorus is the limiting nutrient for phytoplankton. In lakes with unaltered and undisturbed watersheds, year-to-year algae seasonal abundance patterns are fairly consistent. In infertile lakes, zooplankton can effectively graze down populations of phytoplankton. Algae reproduce asexually, primarily by cell division or formation of spores. Algae don't live as long as many other plants because they tend to sink to the lake bottom where light is unavailable for survival. Even the algae that have adaptations to reduce sinking rates, such as gas vacuoles within their cells, rely on high rates of reproduction to maintain their lake presence. To estimate the abundance of algae, limnologists measure chlorophyll *a* pigment concentrations in the upper waters of a lake's pelagic zone. Lakes with average summer water chlorophyll *a* concentrations less than ten micrograms per liter (or parts per billion) generally have high water transparency and low frequency of nuisance algal blooms in the summer. Lakes that have chlorophyll *a* concentrations above twenty to thirty micrograms per liter generally have poor water transparency.

Aquatic plants are a natural part of lake ecosystems and provide benefits to fish, wildlife, and people. Insects, snails, and other invertebrates thrive in aquatic plant stands. Sunfish eat aquatic plants in addition to insects and crustaceans. Bass (*Micropterus*), sunfish (*Lepomis*), and yellow perch (*Perca flavescens*) usually nest in areas with vegetation. In the spring, northern pike (*Esox lucius*) lay their eggs on the remnants of last year's vegetation. Plants then provide shelter for young fish. Some aquatic plants, such as bulrushes, can break down polluting chemicals. Aquatic plants produce oxygen while they absorb phosphorus, nitrogen, and other nutrients, leading to increased water clarity. Eliminating submerged aquatic plants often results in an explosion of algae, followed by a decrease in water clarity. Aquatic plants dampen the force of waves and help prevent shoreline erosion. Plants also stabilize lake bottom sediments. Many aquatic plants produce seeds or tubers that are eaten by wildlife. Bulrushes, pondweeds, and wild rice (*Zizania palustris*) are important foods for ducks. Aquatic plants also provide habitat for loons (*Gavia*), grebes (Podicipedidae), herons (Ardeidae), muskrats (*Ondatra zibethicus*), and other wildlife.

The aesthetic attraction of a lake is defined to a large extent by the vegetation in and around the lake. Plants such as water lilies (Nymphaeaceae), arrowhead (*Sagittaria*), and pickerel-weed (*Pontederia*) have striking leaves or showy flowers. Many aquatic plant species show remarkable differences in their physical characteristics depending on where they are growing (polymorphism). For example, arrowhead has its namesake arrow-shaped leaf while growing alongshore but looks like a different plant, with a ribbon-type leaf, when growing in deeper water. Limnologists and botanists group aquatic plants by their dominant life form.

Submersed plants are part of a diverse group that includes macroalgae (*Chara* and *Nitella*), primitive plants (quillwort, *Isoetes,* and mosses), and a wide array of vascular plants from numerous plant families. Vascular submersed plants often have thin leaves, limp structure (as they have no lignin within their cell walls), and no stomata on the leaves. Because these plants are supported by and soaking in water, their vascular systems are simple, with few conducting vessels in the stem. Submersed plants commonly express one of two leaf types: dissected leaves or, less commonly, entire leaves. In both types, submersed plants have high leaf surface area to plant volume ratios to maximize efficiency in nutrient uptake and underwater photosynthesis.

To gain appreciation for submersed plants in our lakes it helps to scuba dive in and around these underwater forests. The places these plants create are amazing. One can see young bluegill (*Lepomis macrochirus*) and bass hovering in forest openings, schools of bluntnose minnows (*Pimephales notatus*) darting around, and yellow perch appearing confused. Look closely and you will see caddis fly larvae grazing and dragonfly nymphs hunting on the "trees." In the understory, Iowa darters (*Etheostoma exile*) scurry to catch insects. The number of different plants and animals one sees in a short underwater adventure will likely surprise even the most seasoned lakeshore resident. It is a world we spend little time observing.

Some of the most common submersed plants found in lakes include muskgrass, pondweeds, coontail, and milfoil. Muskgrass (*Chara*) is common in many hard-water lakes. Muskgrass has a musky odor and a brittle texture due to calcium in the water depositing on the plant. This plant resembles a vascular plant but is a macroalgae and therefore lacks leaves, stems, roots, and flowers. It often carpets the lake bottom, where it provides important habitat for fish spawning and cover for vulnerable species. Muskgrass grows on a variety of lake bottoms, and it is often the first species to colonize open areas of lake bottom where it helps to stabilize the sediment.

Coontail is one of the most common submersed flowering plants. This free-floating plant lacks roots, so it sometimes drifts around the lake. It has dissected leaves that are serrated and whorl around the stem. Coontail provides important fish habitat, and its stands are home to many species of aquatic insects. Northern water milfoil (*Myriophyllum sibiricum*) is similar to coontail, but its leaves are structured like feathers whorled on the stem. Water milfoil is a rooted, perennial plant whose stems and leaves may reach the water surface, particularly in depths less than ten feet. Its flower stalks emerge above water. This plant spreads primarily by stem fragments, overwinters by hardy rootstalks, and reproduces by winter buds, which are essentially packed leaves that form in the fall at the branch tips. Northern water milfoil is not tolerant of turbidity and grows best in clear-water lakes. Like coontail, this plant provides fish shelter and insect habitat. Broad-leaf pondweeds (*Potamogeton*) are the most stunning of the submersed plants, with green to reddish-brown leaves and prominent midveins. This group includes largeleaf pondweed Illinois pondweed (*P. illinoensis*), white-stem pondweed (*P. praelongus*), and clasping-leaf pondweed (*P. richardsonii*). These plants may form floating leaves, and their flowers and seed heads extend out of the water. Many fish species use these plants for cover and resting areas.

Floating-leaf plants are rooted on the lake bottom, and their leaves and flowers float on the water surface. Water lilies are a well-known example, and many people admire their beauty and durability. This plant uses a variety of traits to survive and prosper. First, the leaves are often oval or circular in shape, unlobed, strong, leathery, resistant to tearing, buoyant due to air chambers, and with a waxy cuticle to repel water. Second, the leaf petioles and flower stems are remarkably strong, enough that they are able to hold on to the leaves and flowers in windy conditions. Finally, the leaf stomata are located only on the upper leaf surface, as this is where gas exchange is most productive. Floating-leaf plants are found in shallow areas and in places protected from high winds. White and yellow water lilies are found in many lakes. White water lily (*Nymphaea odorata*) has showy white flowers with a yellow center and round large leaves (six to twelve inches in diameter) with radiating veins. The leaves have a slit from the edge to the center where the stem is attached. The stem has air channels to transport oxygen to the roots. The large (two- to six-inch) flowers float, have their own stems, and open their petals for only several hours in the late morning and early afternoon. Yellow water lily (*Nuphar variegata*) has a small ball-shaped yellow flower and heart-shaped leaves with parallel veins. White and yellow water lilies often co-occur, but yellow water lily is generally found in shallower water than white water lily. Moose love to eat water lilies, and beaver will uproot plants to eat the starch-rich roots. You may be able to follow the foraging patterns of the nocturnal beaver when you see the thick, long, scarred, brownish water lily roots floating at the surface. Water shield (*Brasenia schreberi*) has small (two- to five-inch) oval-shaped leaves with no slit and small reddish or dull-purple flowers. The leaves are green on the upper surface, while the underside and stems are reddish-purple. The leaves and stems of water shield have a slippery, gelatinous coating. Whereas water lilies grow in silty or mucky shallow areas, water shield is more common in clear, low alkalinity lakes. These beautiful plants colonize by expanding from roots and by seed.

©ANDREA LEE LAMBRECHT

WHITE WATER LILY FLOWERING ALONGSHORE.

Emergent plants are rooted in the lake bottom, and their leaves or stems extend out of the shallow water. There are annual and perennial emergent plant species. Emergent plants have thick cell walls for added rigidity, stem tissue that is water impervious and gas porous, young leaves that can respire underwater until they emerge, and roots and rhizomes that can endure periods of low oxygen. Emergent plants are robust enough to hold the plant against wind and waves. Many emergent plants flower, though their flowers are generally small. Emergent vegetation is critical for numerous fish and wildlife species. Amphibians, ducks, loons, herons, and other wildlife depend on emergent vegetation stands for feeding, breeding, nesting, and shelter. Emergent vegetation provides fish with foraging areas and refuge from predators. Many fish depend on this habitat for at least part of their life cycle. Common emergent plants include wild rice and bulrush.

Wild rice is an annual grass. Wild rice is only distantly related to the domesticated common rice (*Oryza sativa*) but is an important food source for people and for lake waterfowl. Wild rice, which reaches a height of three to six feet, grows in shallow water in lakes, ponds, and slow-flowing streams. In early summer, wild rice leaves are floating on the lake surface. By midsummer the stems and leaves emerge and they quickly flower. Flowers are long (up to two feet) with male spikelets at the bottom and one-flowered female spikelets at the top. Mature wild rice stems are hollow but quite fibrous; anyone who boats into these plant stands will quickly find their propeller entangled. The seeds of wild rice mature in late summer and soon fall into the water. If they do not freeze or dry out, they will germinate in the spring.

Hardstem bulrush (*Schoenoplectus acutus*) and softstem bulrush (*Schoenoplectus tabernaemontani*) are widely distributed emergent plants that provide fish spawning habitat, shelter for eggs and juvenile amphibians and fish, colonization sites for aquatic invertebrates, and protection from shore erosion by dampening wave energy. They are perennial plants with round stems and no leaves (if present, only found at the base of the plant). They grow in shallow water where they average about three to four feet in height; however, they can also be giant plants in deep, clear-water locations (hardstem bulrush) or in mucky, quiet waters (softstem bulrush). In midsummer and on fertile plants, clusters of small flowers form near the end of the stem. Other bulrush species living around lakes include wool grass (*S. cyperinus*), three-square (*S. pungens*), and river bulrush (*Bolboschoenus fluviatalis*). Bulrushes grow in colonies or stands, sometimes by themselves but often with other aquatic plants. Bulrush stands can expand by seed or by sending out thick, brown rhizomes. These stands are vulnerable to shoreline activities as well as grazing by muskrats and other animals.

There are several other interesting emergent plants. Arrowhead (*Sagittaria*) is a perennial plant that can exist as an emergent, floating-leaf, or submersed plant, but it is most recognizable as an emergent with a rosette of dark green arrowhead-shaped leaves. Arrowhead in bloom has beautiful white flowers, and fertilized flowers grow into green seed heads. It grows in rich soils and provides food for beavers and muskrats. Spikerush (*Eleocharis*) resembles the closely related bulrushes. Like bulrush, spikerush has hollow stems, but this plant tends to be smaller than bulrush. Spikerush lacks leaves (remnant leaves only exist as sheaths at the stem base), and in summer the stems are topped with brown spikelet flower heads. This genera is rich in species, including needle spikerush (*E. acicularius*) with stems less than six inches tall, squarestem spikerush (*E. quadrangulata*) with distinctive squarish stems, and common spikerush (*E. palustris*) that grows in mat-like stands. Horsetail (*Equisetum fluviatile*) is a unique plant. It is a primitive perennial plant with jointed, hollow stems. At the stem joints, whorled, black-tipped scales are present. The plant reaches an average height of three feet or

© Andrea Lee Lambrecht

WILD RICE AND YELLOW WATER LILY PROVIDING COVER TO FISH ALONGSHORE.

less. Horsetail stems are rough and have a high concentration of silica. At one time people used horsetail stems for scouring or polishing, and the plant was historically called scouring-rush. Horsetail reproduces vegetatively by rhizomes and by spores, which are produced on a cone at the tip of the stem. Horsetail prefers sandy soils within the littoral zone. Bur reed (*Sparganium*) is a perennial plant in the cattail family. It has flowers in spherical bur-like clusters on stalks, and when mature it has green or brown achene fruits. The leaf blades, which are about one inch wide, are thick and bright green. Bur reed can reach a height of three feet. Nine species of bur reed occur in North American lakes, some of which are floating-leaf plants. The suite of emergent plants will vary for each lake and may include plants like blue flag iris (*Iris versicolor*), swamp milkweed (*Asclepias incarnata*), pickerelweed (*Pontederia cordata*), water plantain (*Alisma*), and cattail (*Typha*).

Upland plants growing along the shore are also critical factors in determining the lake's ecological integrity. The most effective approach to lake protection requires intact shoreline buffers of diverse vegetation. By slowing the movement of rainwater runoff, buffer vegetation allows sediment contained in the stormwater to settle out. Pollutant removal increases with increasing buffer width. Vegetative buffers less than fifty feet wide are generally inadequate to provide long-term water-quality protection. A fifty-foot buffer may remove about 65 to

70 percent of the phosphorus runoff. Substantially greater widths of natural vegetation are needed to further improve removal rates. Effectiveness of shoreline buffers for phosphorus removal is a function of width and slope. In low slope areas, fifty-foot shoreline buffers appear to be sufficient, but as slope increases, buffer widths of one hundred feet or greater are warranted. Shoreline vegetation also reduces bank and shoreline erosion; a natural shoreline has plants with deep roots that hold the shoreline together. For lakes, trees are valuable in many ways. While on shore they shade, screen, and provide wildlife habitat, and in death they provide cover and habitat. From small branches to whole trees, wood can provide simple habitat for aquatic insects to intricate hiding places for fish.

Shoreline buffers also provide other benefits. A shoreline buffer, as viewed from the cabin, provides a beautiful picture frame for the lake. With judicious pruning of branches of trees and shrubs within the buffer, this naturally vegetated space screens the view of the neighbors yet provides a view out onto the lake. A forested shoreline buffer also protects the privacy of your cabin, shielding it from people recreating out on the lake.

ANIMALS

The diversity of animals that depend on our lakes is broad. The animal food web of a lake is a tangled and chaotic one that morphs with the seasons and changes with the addition and subtraction of species. Bacteria benefit the most from the subtractions; as species die, more energy settles to the lake bottom for their consumption. We benefit the most from the additions and from the resilience of a robust, complex web.

We should all give thanks to bacteria for their service in transforming plant and animal matter back to basic gaseous molecules and compounds, mostly carbon dioxide and methane. If we had to pay to clean up the dead plants and animals in a lake, it would cost a fortune. Algae settle to the bottom, pondweeds fall apart, and fish turn belly-up; luckily the bacteria and fungi are working for free. Lake sediments are the primary location for microbial activity (several orders of magnitude larger than in the water above). If a system has large inputs of organic matter that accumulates faster than it decomposes, the lake bottom becomes anaerobic. Rapid sedimentation of organic matter occurs, creating thick deposits of muck. Each year a new layer of lake sediment is formed from the silt and sediment that settle out of the water and from the dead plant and animal tissue that aren't consumed or broken down into gases by bacteria. Bacteria are also found in the open water, their abundance often correlated with phytoplankton abundance. These planktonic bacteria decompose dead algae and zooplankton. Littoral zones also have measurable levels of microbial activity. Here, both aquatic plants and land sources of organic matter are available for bacterial growth.

Lake zooplankton include protozoa, rotifers, and the microcrustaceans cladocerans and copepods. Protozoans are normally a small component of zooplankton, but they can become abundant when planktonic bacteria, their primary food item, become abundant. Rotifers are often the most abundant zooplankton organism. This group is comprised of many species; *Keratella* is present in many North American lakes. Rotifers are odd creatures, with a worm-like body shape and sometimes a funnel-like appendage surrounding the mouth. Rotifers prey on bacteria, protozoans, and small algae. Unlike rotifers, cladocerans and copepods look more familiar, though like the other zooplankton their bodies are transparent. A cladoceran has a head, body, legs, compound eyes, and antennae that have morphed into swimming append-ages. Cladocerans include the ubiquitous *Daphnia* species. Copepods look like the back end of miniature shrimp with added antennae. Cladocerans and copepods are filter feeders, for-aging or grazing on planktonic algae. It's hardly a carefree existence for these zooplankton. Cisco, perch, and bluegill like to pick out the big ones, slurp them in, and then sieve them out of the water with their gill rakers. Many cladocerans feed only in the predator-rich surface waters at night when they are less vulnerable to predation.

Aquatic invertebrates represent many animal phyla, including freshwater sponges, nema-todes, leeches, insects, and the crustacean orders of mysids, ostrapods, isopods, amphipods, and crayfish. Crayfish are a common animal in many lakes. They prefer hard lake bottom areas, and they are omnivores, feeding on plants, a full range of invertebrates, and occasionally small fish and fish eggs. Other highly visible invertebrates include snails and mussels. Snails are herbivores and use their sclerotized jaws to scrape algae from rock and plant surfaces. Mus-sels are filter feeders, straining particles from the water by using a siphon. They are long-lived, and may even outlive us. Mussels depend on fish to raise their offspring. Mussel larvae, called glochidia, parasitically attach to fish gills or skin. After the larvae mature into juvenile mussels they release themselves and sink to the lake bottom to grow into adults.

Insects are the most common animals on the planet, and the same holds true in and around the lake. Although most insects live on land, there are many aquatic insect orders including true bugs, dragonflies, stoneflies, mayflies, true flies, caddis flies, alderflies, and beetles. In lakes, the number of insect species and the amount of insect biomass are

©ROBERT FITZSIMMONS

FEMALE CRAYFISH WITH INCUBATING EGGS ATTACHED TO HER ABDOMEN.

impressive. Midges, or true flies, can be found almost everywhere dissolved oxygen is present at the sediment–water interface, and even a few places where it's not. Depending on the species, nonbiting larvae of the chironomid fly family live for several weeks to a year on the lake bottom. The phantom midge (*Chaoborus*) is common in North American lakes and unlike many other midges it does not live on the lake bottom, instead it is a zooplankton that preys on copepods and cladocerans. Midge larvae metamorphose in the water and emerge from the water as adults, living only long enough to mate and lay eggs back into the lake. Several of the true bugs are interesting to watch move in the water. Water boatmen (*Corixidae*) have elongated, oar-like legs and swim dorsal side up, unlike their relatives the backswimmers (*Notonectidae*) that swim upside-down. Water striders (*Gerridae*) run, hunt, and rest on the water surface by using long legs and fine water-repellant hairs to take advantage of water surface tension.

There are over 150 species of dragonflies and damselflies that inhabit North American lakes. You can tell a dragonfly from a damselfly by the way the wings are held at rest: dragonflies hold their wings horizontally outward, and damselflies hold them folded and generally upward. These insects are such beautiful creatures in flight. They flit and hover with grace. But dragonflies are aquatic insects, which means they also have a nymphal life stage.

Dragonfly adults are found flying on sunny, warm days near almost any water body where their nymphs live. Although some dragonflies only fly during bright sunlight, others are out during dawn or dusk. The adult dragonfly hunts by catching flying mosquitoes and other small insects with its spiny forelegs. Dragonflies can fly as fast as twenty miles per hour, and they use their two large compound eyes to survey their surroundings. For some species, males defend a territory over a stretch of water where a female will lay eggs, and they patrol this area by flying back and forth. Dragonflies lay their eggs a number of ways. In some species, the female flies above the water and scatters her eggs by touching the tip of her abdomen to the surface. Some deposit their eggs in or on submerged or emergent plants. The number of eggs laid by a female varies from a few hundred to a few thousand. Most dragonfly eggs hatch in twelve to thirty days, but some species go through the winter in the egg stage.

Dragonfly nymphs are unique looking, robust animals with long legs. They are generally less than one inch long and can be green, grey, or brown colored. One of their most striking features is their lower lip, which is greatly modified to serve as a food-snatching device. In addition, nymphs can swim by jet propulsion by squirting water out from the ends of their abdomens. Dragonfly nymphs are commonly found on submerged vegetation in the shallows of lakes and ponds and are rare in polluted water. A large part of the diet for many dragonfly species, both nymphs and adults, is mosquitoes. Nymphs also eat other aquatic insects. The nymphs themselves are food for birds, fish, and other large aquatic insects. Depending on the species, nymphs spend a few months to three years in the water before they emerge to become adults. When the time is right, the nymphs crawl out of the water, usually on emergent vegetation. The actual process of emergence occurs when the transformed nymph crawls out of its exoskeleton skin. Before flying off, the newly emerged adult clings to the nymph skeleton for an hour or so while the wings become dry and stiff.

The largest dragonfly of North American lakes is the common green darner (*Anax junius*). This species attains an adult length over three inches and a wingspan greater than four inches. It has a green thorax and a bright blue (male) or brown-purple (female) abdomen. The common green darner is one of several dragonfly species that migrates in spring and fall. Similar to the grand migration of monarch butterflies to Mexico, common green darners, in groups of a few to masses of millions, migrate south for the winter.

Fish are certainly the best-known lake creatures, and where in the lake they are found depends on many factors. First, each species of fish has specific water temperature, oxygen concentration, and light preferences, which may vary with age or size. Second, some species have affinities for structure (e.g., aquatic vegetation, downed trees) or lake bottom. Third, at the daily and seasonal scale, habitat use can be a function of spawning requirements, feeding needs, predator avoidance behaviors, and a host of fish and other organism interactions. Walleye, a species of the eastern United States and much of Canada, provides a good demonstration of factors influencing habitat use. This species's habitat use is strongly influenced by water temperature and light intensity. In the summer, since many lakes are thermally stratified, walleye generally seek out parts of the lake where the water temperature is within its preferred range, 64–72F (18–22C), and with sufficient dissolved oxygen (more than six milligrams per liter). Walleye evolved to hunt at night to minimize competition with other predators and to maximize success in capturing prey. The primary adaptation is a layer in the eye that reflects light back through the retina called a tapetum lucidum, which increases light reaching the photoreceptors of the eye. (This adaptation evolved independently in several animals over three hundred million years ago.) This adaption allows walleye to see in very low light conditions, as compared to their primary prey, yellow perch, which do not possess any specialized light-gathering eye tissue. During the day walleye will avoid well-lit shallow areas; rather they will be in deep water either moving slowly near bottom or resting in contact with the lake bottom, as is their affinity. Other fish species have no or less lake bottom affinity. Cisco and lake whitefish will suspend in the water column and migrate within that water column during the day, rising in the night to feed and moving down with the sunrise. Walleye move to deeper water around fall turnover, and they generally stay in deep water until spring when mature fish move to shallow water for spawning. In their first year of life young walleye habitat preferences change as they grow. After hatching, walleye fry are pelagic, feeding on zooplankton; a month later young walleye, just inches long, move back to nearshore shallow water to feed on young yellow perch; and by the end of summer, when they about six inches, they move to deeper water.

Besides sunfish, bass, trout, and pike, there are many other fish species that live in lakes, and it is not uncommon to find over fifty species in a lake. Small, often overlooked fish species include bluntnose minnow, spottail shiner *(Notropis hudsonius)*, emerald shiner *(N. atherinoides)*, mimic shiner *(N. volucellus)*, longnose dace *(Rhinichthys cataractae)*, Iowa darter, johnny darter *(Etheostoma nigrum)*, and logperch *(Percina caprodes)*. Natural resource management agencies use the presence and abundance of fish and other animal species to assess environmental quality. Certain small fish are indicators of the health of nearshore habitat. Pugnose shiner *(Notropis anogenus)*, least darter *(Etheostoma microperca)*, longear sunfish *(Lepomis megalotis)*, blackchin shiner *(N. heterodon)*, blacknose shiner *(N. heterolepis)*, and banded killifish *(Fundulus diaphanus)* are associated with large, nearshore stands of muskgrass or other aquatic plants, and they are sensitive to disturbance. They have been extirpated from lakes where extensive watershed and lakeshore development has occurred.

Northern pike are one of the most common game fish in North American. Adult northern pike move into the shallows as the ice starts to break up, looking for suitable spawning habitat. They seek out areas that had emergent plants, like bulrushes, grasses, sedges, and cattails, growing the previous summer. Here the female pike will deposit the eggs, and the male (or males) will disperse sperm to fertilize them. Depending on her size, a female northern pike lays between eight thousand and five hundred thousand eggs. Why the large number? It is important for northern pike (and most fish species) to produce large numbers of eggs for two reasons. First, most eggs do not survive. Either the conditions are hostile or the location is unfavorable, despite efforts by the parent to find suitable habitat. Second, high survival years are critical to sustain the pike population. High survival years occur when weather, environmental conditions, and prey and predator populations are favorable, and these conditions are infrequent and irregular. Many fish populations have interesting demographics.

The northern pike eggs are small and amber-colored, and they adhere to the submerged remnants of vegetation if they are fortunate to land on these surfaces that provide good conditions for hatching. Eggs that settle into the lake's mucky bottom will likely die due to poor oxygen conditions. Eggs hatch in twelve to fourteen days. In their first few days as fry, northern pike have no developed mouth. They hide in the vegetation or near the lake bottom. At this stage, they can adhere to vegetation using a sucker-like organ on the top of the head. After absorption of the yolk sac, a young pike's survival depends on plankton. If plankton are scarce or of the wrong size, then northern pike growth and survival will be poor. By early summer, juvenile pike are eating aquatic insects and small fish, such as darters, yellow perch, and minnows. Young northern pike are highly selective in where they hang out, and you would be too if someone was fixing to make you their next meal. Young northern pike might choose to remain in the shallow-water marshes where they hatched, or they might live in vegetation found in the shallow-water, nearshore areas around the lake itself. Vegetation provides important refuge from predators, such as walleye, bass, yellow perch, and even adult northern pike. It is a rough lake out there for a northern pike, and survival of the fittest rules. By the end of summer, a young pike may measure six inches in length, but it will be another two to six years before it reaches maturity. Pike are excellent predators, concealing themselves in aquatic vegetation and darting out to capture fish and other prey. Pike can reach lengths of four feet or more and can live up to twenty-five years.

Bullfrogs (*Lithobates catesbeianus*), green frogs (*L. clamitans*), and mink frogs (*L. septentrionalis*) are shoreline-dependent species that inhabit many North American lakes. They establish and defend distinct territories and tend to remain along the shores of lakes or in areas of shallow water with emergent vegetation throughout the summer breeding season. To determine if you have these frogs at your lake, just listen. Male bullfrogs have a loud low-pitched two-part bellow. Male green frogs have a call similar to the plucking of a banjo string, and male mink frogs make a "knock-knock-knock" sounding call. The frogs call during the breeding season, which for bullfrogs begins in late spring. Green and mink frogs begin their calling in late May and can often be heard until early August. The bullfrog is the largest frog in North America, averaging about six inches in length, and it is typically dark olive to pale green in color. Green frogs are medium-sized, greenish or brownish frogs, with small dark spots on the back, a light-colored belly, and a large tympanum (eardrum). They can be found in a variety of habitats surrounding lakes, streams, marshes, and swamps but are strongly associated with the shallow water of lakes and the shoreline. Mink frogs are typically green in color with darker green or brown mottling. Mink frogs inhabit quiet waters near the edges of wooded lakes, ponds, and streams and are considered the most aquatic of the frogs. If you move quietly along the edge of a water lily bed, you may observe mink frogs sitting on the pads. The presence of frogs is a good sign and is often used as an indicator of habitat quality.

If there are frogs on the lake, there are probably turtles as well. The presence of turtles, like frogs, is a sign of good environmental quality, and their decline or disappearance locally is a red flag. Snapping (*Chelydra serpentina*) and painted (*C. picta*) turtles are the two common turtle species inhabiting central North American lakes. Snapping turtles get big, and it is advisable to give them some personal space. Painted turtles are more tolerant. Today, painted turtles are considered one of the most successful lakeshore animal species. They have existed for fifteen million years, are widely distributed across North America, and are abundant in their preferred habitat. Painted turtles are small turtles with colorful shells. The bottom of the shell often has black and yellow patterns with red margins. Both snapping turtles and painted turtles need basking areas, such as rocks or logs extending above the water surface, where they can sit and absorb the warmth of the sun. They also require areas with extensive mud bottoms and abundant aquatic vegetation. Turtles are omnivores, eating aquatic plants, insects, crayfish, and fish. The snapping and painted turtles you see on land are often females, preparing in early summer to lay their eggs in sand. The turtle nest sits unattended through the summer, and the young hatch in late summer or early fall.

If you take an early morning summer walk to a lake, and you walk through a mix of habitat from upland forests to fringe wetlands, it is conceivable that you could hear over 150 different species of birds. From the common loon's (*Gavia immer*) distinctive call to the slow monotone trill of the swamp sparrow (*Melospiza georgiana*) to the buzzy, sharp, high-pitched notes of the eastern kingbird (*Tyrannus tyrannus*), the shoreline is awash with bird songs. Spring and fall bird migration periods have higher diversities of birds alongshore than other times of the year. The majority of the more than 650 bird species that nest in Canada and the United States winter in the neotropics, and we enjoy the color, sound, and energy they bring as they migrate. Shorelines are important bird migratory habitat.

©ROBERT FITZSIMMONS

A PAINTED TURTLE SWIMMING ABOVE THE SUBMERGED VEGETATION.

© ANDREA LEE LAMBRECHT

EASTERN KINGBIRD HARVESTING THE HAIRY SEEDS OF A CATTAIL FOR ITS NEST.

© ANDREA LEE LAMBRECHT

KILLDEER HUNTING INSECTS ON THE BEACH.

Stopover sites and migration corridors are vital to migrating birds for resting, feeding, and travel efficiency. In spring, adult midges emerge from lakes and provide an abundant food source onshore for hungry migrating warblers and other bird insectivores heading north. In fall, seeds from sago pondweed (*Stuckenia pectinata*) and wild rice, as well as the seeds, foliage, and tubers from other aquatic plants, attract and support waterfowl moving south. Large lakes also influence bird migration as wide expanses of open water create physical or emotional challenges to some species. The western shore of Lake Superior is a critical flyway for a large number of hawks, falcons, and eagles. The Lake Michigan flyway provides a corridor for hundreds of bird species, including a vast array of warblers, sparrows, flycatchers, swallows, and shorebirds. The other Great Lakes also have areas that funnel birds alongshore.

The common loon is a large piscivorous waterbird of northern latitudes. It winters along the coasts of the Atlantic and Pacific Oceans and the Gulf of Mexico. The breeding range is primarily in Canada, but a small proportion of the population summers in the northern United States. Loons commence their diurnal migration in the morning. Spring migrations generally occur rapidly, and the presence of a loon on a breeding lake occurs quickly after lake ice-out. The fall loon migration period is often more prolonged as loons use large lakes along the way to rest and feed. Not surprisingly breeding populations differ in their wintering areas. Loons coming from their summer lakes in the Canadian eastern provinces, Maine, and New Hampshire migrate to the Atlantic coast and overwinter in the Gulf of Maine. Loons from Quebec, New York, and Vermont might migrate to the coast either around Long Island Sound south to New Jersey or within Chesapeake Bay. Loons summering in northern Ontario may stop over on Lake Ontario before heading to the Atlantic coast, while loons summering in the Great Lakes region may stop at one or more of the Great Lakes. Some of these Great Lakes region loons may head south from Lake Michigan resting on southern U.S. lakes before finally reaching their Gulf of Mexico wintering area, while others move easterly to the coast. Loons summering in the midcontinent, from the far north to Montana, may stop over in lakes east of the Sierra Nevada mountain range (common stopover lakes include Walker and Pyramid Lakes, Lake Tahoe, and Lake Mead), and later they head west to winter on the coast of California or in the Gulf of California. Loons breeding in Alaska and western Canada winter on the Pacific coast. Biologists have used radio and satellite transmitters to better understand loon migration behavior and timing.

For northern lakes, loons are a good indicator species for water clarity and ecosystem health. Loons seek breeding habitat in clear-water lakes with adequate small fish abundances. Since loons rely on sight to locate and capture their prey, they are dependent on high water transparency. Loons use their large webbed feet to hunt by diving and swimming after their prey, which include yellow perch, many species from the minnow family, and a variety of young or small fish from other fish families (trout, sucker, sunfish, etc.). They also eat aquatic insects, crayfish, amphipods, leeches, frogs, and salamanders. Suitable small lakes often have a single mated pair of loons, and large lakes can have numerous pairs nesting in secluded bays and on small islands.

Loons face many threats to their population growth, breeding range, and reproductive success. The environmental stressors include disturbances to coastal marine areas, alteration of nesting habitat, and habitat shifts due to climate change, pollution (e.g., mercury deposition and lead fishing tackle), and eutrophication. The southern range of the common loon's breeding range has shifted north, perhaps related to these and other environmental alterations caused by human development and activities. Human activities that disturb loon nesting are a serious concern, as their nesting requirements are very specific. Although excellent swimmers, loons are poor walkers because their feet are set far back on the body. Therefore, they nest near the shore or on floating mats of vegetation. Loon nests are generally found in areas protected from prevailing winds and with good views to their open-water territory. Male loons select nesting sites and aggressively defend their territories. They show high nest site fidelity for successful nesting sites. Although loons are somewhat adaptable to human activities, efforts by lakeshore residents and lake users to protect loons' nesting habitat and lake water quality will help ensure the long-term success of the species.

Many mammal species inhabit lake shoreland. Each of us has favorites that delight us by their presence, sometimes only with a fleeting or flitting glimpse. Observations of large animals like moose, bear, deer, coyotes, or fox are appreciated even if some of us have some apprehension about their presence due to the lack of understanding of their motives. Most of our biophilia is directed to species we can relate to, like those similar to our domesticated animals. However, common small mammals are often important for ensuring ecosystem resilience and integrity. Across North American lakes, the muskrat is a good indicator for healthy wetlands and shoreline vegetation. This rabbit-sized, water-dependent rodent feeds on aquatic plants. In lakes, muskrats use bank burrows with underwater entrances, and occasionally they will build small houses of vegetation in shallow-water areas. Many animals prey on muskrat, including mink (*Neovison vison*), fox, coyotes, wolves, bobcats, bears, owls, and hawks. Muskrats are mostly nocturnal, a strategy to avoid predators, but during low light conditions you may observe individuals swimming nearshore in search of food or during the day sunning on a log or sitting along a muddy shoreline neatly eating pieces of water lily, cattail, bur reed, or bulrush vegetation. They are the primary herbivore within lake wetland fringes, and abundant muskrat populations control the expansion of cattail stands and influence wetland plant composition and function. Muskrat herbivory increases plant diversity in these wetlands.

And then there are humans, undeniably part of the ecosystem. We investigate human impacts and opportunities in later chapters. Before that, we need to know a few things about the way various parts of the lake landscape connect.

CHAPTER 2

Lake Ecology

ECOLOGY IS THE STUDY OF THE RELATIONSHIP OF ORGANISMS TO ONE OTHER AND TO their physical environment. Natural selection, predator–prey interactions, and population dynamics are all included within the framework of ecology. Like other disciplines of biology, ecology is based on the fundamental science of physics. Within ecosystems, the laws of science, such as the thermodynamic laws are often hidden to the casual observer by the complexity of species interactions and the environment. The first thermodynamic law states that energy in an isolated system is conserved, and the second law states that energy disperses in an isolated system. As with theories of physics, most theories in ecology have been around for a long time. In 1942, Raymond Lindeman related the second law of thermodynamics to energy flow in Cedar Bog Lake of central Minnesota and Lake Mendota of Wisconsin. Lindeman used the then newly defined concepts of ecosystems and food webs and tied them to energy flow. Lindeman documented the flow of energy through lake food webs, from primary producers (algae and other plants) to herbivores (zooplankton) to predators (fish and humans). As energy moves through the food web, a large amount is unavailable to organisms. On average, only about 10 percent of the energy from one level of the food web transfers up a level (for example from herbivores to predators). Animals use energy to survive and reproduce; the end result is that there must be substantially fewer predators than prey. The law of conservation of energy states that energy can neither be created nor destroyed. This holds true within ecosystems, where energy is not truly lost, but just lost into the environment, typically in the form of heat that is unavailable to organisms at higher trophic levels. Given this energy dissipation between trophic levels, the overall shape of a food web is a pyramid. It also follows that a food web cannot extend more than a small number of levels. Conversion of food must follow chemical and thermodynamic laws. Species do not exist because of magic. Lindeman's research on lake ecosystems proved this to be true.

These laws of nature, along with the theory of natural selection, population dynamic concepts, community succession, and predator–prey theory, provide the theoretical framework of ecology. Lakes and wetlands follow these general laws of nature, and so should their management. Several principles of nature follow from this framework. Here are four core principles.

Principle #1: The law of the minimum applies to organisms and populations within ecosystems. Every individual and population has space and condition requirements, and the size of that niche influences their fate. While an organism is dependent on a number of different resources, at a single point in time one of those resources will limit its growth—the limiting factor. The scarcest resource determines the rate of growth for an individual or a population. For example, lake algae growth is dependent on carbon, nitrogen, phosphorus, and other essential nutrients, but phosphorus is often the scarcest and its availability therefore generally determines algae abundance. There is always a limiting factor.

Principle #2: Organisms and ecosystems have histories that constrain the future. Through genetic variation (random mutation and recombinations) and natural selection, every organism has evolved from something similar. Evolution is a bottom-up process. An organism may evolve in the future, but in many ways it will be quite similar to its recent evolutionary ancestors. We evolved from earlier hominids several hundred thousand years ago. Our needs are similar to those of our ancestors. If, in the future, another species evolves from us, its needs are likely to be similar to our own. Every ecosystem also had a past with a community of species and physical resources. The character and function of an ecosystem in the future is constrained by its present condition. An undisturbed ecosystem will operate in the future much as it did in the past. Species and ecosystems are not created out of thin air. The future will likely be similar to the past.

Principle #3: Processes and interactions within ecosystems occur at many different scales. From the molecular scale that determines walleye egg survival on the shore of a lake to the walleye population dynamics in a lake to the range and distribution of walleye in North America, the study of walleye ecology is scale dependent. Scale matters.

Principle #4: Many things are connected to many other things. Organisms are dependent on other organisms, and the probabilities of species interactions are dependent on time and space. The dynamics and patterns that we see in ecosystems are the result of multiple causal factors, both biological and physical. Species diversity is important. Diversity of plants and animals is critical in maintaining the health of an ecosystem. Diversity allows an ecosystem to adapt to varying conditions. Areas with higher biodiversity often are more productive and resilient. Species competition for resources is an important ecological factor, but there are also many others. It is often difficult to know why something happened in a complex system. Stuff happens.

LIMNOLOGY

Limnology is the study of inland waters. Scientists in this field study the interactions of physical, chemical, and biotic attributes of lakes and rivers. Of particular value to lakeshore residents and land-use designers, limnologists study the cycling of carbon, phosphorus, and nitrogen in our lakes to understand the consequences of poor shoreland development and pollution regulation policy.

The carbon cycle and the amount of carbon in lakes are important in determining acid buffering capacity and aquatic plant productivity. Carbon dioxide makes up only a small proportion of gas in the atmosphere (it is, however, increasing due to the burning of coal and oil), but it is abundant in lakes due to its high solubility in water. Carbon dioxide also enters the lake from groundwater, organism respiration, and anaerobic decomposition. Once in the lake, carbon dioxide reacts with alkaline earth metals, primarily calcium, to create carbonates and carbonic acid. A lake with a high concentration of carbonates in the water has a high acid buffering capacity and is also referred to as a high alkalinity lake. Water hardness is a function of the concentration of calcium and magnesium salts; hard-water lakes have a high mineral content. Hard-water lakes are often more productive than soft-water lakes. Soft-water lakes usually have low abundance of aquatic plants, and the plants that are present tend to be rosette or mat-forming species such as water lobelia (*Lobelia dortmanna*), quillwort (*Isoetes*), pipewort (*Eriocaulon*), and water bulrush (*Schoenoplectus subterminalis*). Hard-water lakes usually have abundant stands of pondweeds, muskgrass, coontail, water milfoil, bushy pondweed (*Najas*), and water lilies. Many lakes have sufficient quantities of carbonate minerals to buffer the acid rain created by our carbon dioxide and sulfur dioxide emissions. However, many northern lakes have low acid-buffering capacity, and they face serious problems.

Phosphorus is an essential element for plant and animal life and growth. Phosphorus (as phosphate), along with sugars, serves as the backbone of DNA for Earth's life forms. Phosphorus (as triphosphate), along with adenosine, forms a molecule (adenosine triphosphate, ATP) responsible for energy transfer within cells that maintain life. On our planet, phosphorus is the rarest element of those necessary for life. (The other elements of life are carbon, hydrogen, oxygen, nitrogen, and sulfur.) Phosphorus, as a limiting factor, usually governs the amount of algal and aquatic plant growth within a lake. For this reason, limnologists study the role of phosphorus in lake ecology, and pollution regulators desire a full accounting of phosphorus inputs into lake water to identify reasonable means to maintain or improve water quality. Phosphorus naturally has a spotty distribution across North America, but many soils produced from the weathering of igneous rock have abundant concentrations of phosphorus. Full accounting of phosphorus inputs involve stream, groundwater, rainwater runoff, and atmospheric monitoring.

Once within a lake, phosphorus cycles rapidly. Bacteria, algae, and other plankton absorb available phosphorus quickly, release excess, and soon die. Aquatic plants absorb phosphorus from the water through their leaves and from the sediment by their roots or rhizomes. As submersed plants die, phosphorus is released back to the water. While there is a net movement of phosphorus to lake sediment, the release of phosphorus back into the water is of critical importance in its recycling within many lakes. When the dissolved oxygen concentration of the water declines to zero from bacterial decomposition, this creates a chemical condition that releases phosphorus from the lake sediment. In deep lakes, turnover events that reoxygenate the water send much of the phosphorus back to the sediment as iron in the water reacts with the phosphorus to form a precipitate that settles to the lake bottom.

Total phosphorus concentration in lake water varies by lake. There is a positive relationship between phosphorus concentrations and algal productivity. When total phosphorus concentrations exceed twenty micrograms per liter the amount and severity of algal blooms increase greatly. When phosphorus concentrations reach thirty micrograms per liter, nuisance algal blooms become more frequent. The variability of phosphorus concentration levels in lakes is also a function of different levels of pollution loading across North America. Large stratified lakes are less sensitive to water clarity changes due to phosphorus pollution. Steady phosphorus loading is necessary to sustain high algae productivity in these lakes that initially had low nutrient concentrations. Drainage lakes with high flushing rates can recover if phosphorus pollution is reduced. Shallow lakes, on the other hand, are very vulnerable to phosphorus pollution as phosphorus cycles between the lake sediment and the water due to frequent mixing and sediment resuspension. Because the phosphorus loading in these lakes is internal, reducing external sources of phosphorus loading may not result in water quality restoration.

Nitrogen cycling is also important in lake ecology. Nitrogen is abundant in our atmosphere and can be found in various forms in our lakes. Although phosphorus is often the limiting nutrient in lakes for algae production, nitrogen can also act as a limiting factor (particularly in coastal marine waters). Excess loading of nitrogen can lead to increased eutrophication and degraded water quality. In addition, increased loading of nitrogen has important direct effects on aquatic plant communities. In embayments and nearshore areas of many developed lakes, nitrogen pollution can lead to declines in diversity and structure of aquatic plant communities, with consequences including increased algal growth, loss of important fish habitat, and sediment destabilization. Nitrogen stable-isotope analysis is a powerful tool used to identify anthropogenic nutrient inputs into aquatic systems. Stable isotopes are nonradioactive forms of an element that differ in the number of neutrons; that is, they only differ in mass. Nitrogen has two stable isotopes, nitrogen-14 and nitrogen-15. These occur with a natural abundance of 99.63 percent and 0.37 percent, respectively. Because they are so different in abundance, and because the many biological and geophysical processes in ecosystems discriminate between them, distinct isotopic signatures can be identified and used to trace sources of nitrogen. Within a food web, top predators have the highest nitrogen-15 values, and human wastewater typically has elevated nitrogen-15 values compared to unimpacted groundwater. By analyzing the ratio of nitrogen isotopes from aquatic animal and plant material, limnologists can determine the source of the nitrogen used in their growth.

Excessive nutrients flowing into lakes from mismanaged lands and sewage cause serious problems. Ecosystems change with changes in energy or nutrient flow. The composition of tree species in a forest changes due to changes in light levels reaching the forest floor. In a lake, water quality and plant and animal composition change with the accumulation of nutrients (phosphorus and nitrogen). Limnologists classify lakes within a spectrum from low productivity (oligotrophic) to high productivity (eutrophic). Long periods of time (much longer than those needed for forest succession) are often necessary for a lake to age from an oligotrophic state to a eutrophic state. Our lakeshore actions can greatly speed up this nutrient accumulation succession. State and provincial governments regulate and attempt to manage a range of phosphorus and nitrogen discharges and input sources. Wastewater point sources are required to remove phosphorus to the extent possible; phosphorus limitations have been applied to detergents and lawn fertilizers; sewage treatment system standards have been adopted to minimize phosphorus entering lakes; animal feedlot regulations have been developed to reduce nutrient contributions from these intensive animal operations; and riparian buffers are required in some agricultural areas to capture a proportion of nutrient-rich runoff.

SENSITIVE LAKES

While all lakes are sensitive to human activities in the watershed, shoreland, and nearshore areas, shallow lakes and deep coldwater lakes are least resilient to nutrient additions and hydrologic alterations and are most vulnerable to habitat loss. Shallow lakes often have extensive areas less than fifteen feet deep, and in a healthy state usually have abundant aquatic plant communities. Recent research has shown that such water bodies are very susceptible to nutrient loading. Shallow lakes exhibit two alternating stable states. The first state is characterized by clear water, abundant aquatic vegetation, and shallow bays covered with emergent vegetation. This state is desirable for fish and invertebrates and is associated with excellent waterfowl production. The second state has less diversity, very turbid water, and little or no submerged vegetation. Lakes in this state have heavy algal blooms, poor fish communities, and reduced waterfowl production. Shallow lakes can exist for years as either clear or turbid waters. It takes a major perturbation to move from one state to the other.

Across North America the degradation of shallow lakes has been broad-based, cumulative, and persistent. In central and southwest Minnesota, for example, the majority of the lakes do not support aquatic recreational uses. The reasons vary. Some shallow lakes do not support swimming due to phosphorus loading from a watershed source, such as discharge from a wastewater treatment plant. The vast majority of shallow lakes in southwest and northwest Minnesota and southern Wisconsin and Michigan have highly agricultural watersheds. Runoff from these agricultural lands is typically very high in phosphorus. High nutrient loading from these watersheds into shallow lakes promotes internal recycling of phosphorus and typically leads to high in-lake phosphorus concentrations and subsequently to nuisance blue-green algal blooms and low transparency. The combination of high nutrient loading and the limited assimilative capacity of shallow lakes often limits the degree to which the water quality of these lakes might be improved.

Coldwater lakes with oxygenated deep water are also highly sensitive to the consequences of development. These lakes provide habitat for lake trout, lake whitefish, and cisco. Eutrophication threatens these fish and their habitat. Increased fertility, algal production, and sedimentation result in increased aerobic decomposition of organic matter in the deep water and lead to lower oxygen levels. The reduction of the deepwater oxygen level in these lakes forces coolwater fish to move up in the water column, causing them physiological stress and altering the predator–prey dynamic. Global warming is also affecting the thermal conditions of coldwater lakes. They have longer open-water seasons, longer periods of stratification, and deeper extents of warm water; all of these changes threaten coldwater fish. To many lakeshore lovers, the fish communities that include lake trout and other coldwater species define northern lakes, and the loss of these species may redefine where the North begins.

SENSITIVE LAKESHORE

Not all lakeshore is created equal. The shoreline and nearshore areas are critical to the health and well-being of fish and wildlife. Many fish and wildlife species are highly dependent on naturally vegetated shorelines as habitat for feeding, resting, mating, and juvenile life stages. Some lakeshore is more sensitive or vulnerable to human activities. Sensitive areas are places that provide unique or critical ecological habitat, are vulnerable to change, or are exceedingly difficult to restore. Sensitive lakeshore identification is often based on fundamental conservation principles. These principles deal with content, context, heterogeneity, and connectivity.

The biological and physical content, as well as the social structure, determine if a place is sensitive. Rare plants and animals that are vulnerable to development should be identified and mapped. Diversity of both plant and animal species is critical to maintaining the health of an ecosystem. Diversity allows an ecosystem to adapt to varying conditions. Places with many plant and animal species are more productive and resistant to stresses than those with only a few species. Diverse habitats are fundamental in allowing an area to have high plant and animal diversity. Hot spots of diversity are valuable to protect. Unique physical features, such as groundwater recharge areas, bluffs, and productive soils, are also critical content to protect. In addition, places where people enjoy and interact with nature are vital. Littoral areas within lakes provide important habitat and ecosystem services, as do wetlands, bogs, and fens. Wetlands serve a variety of functions. They support a diversity of flora and fauna, many of which are wetland obligates. They also act as a buffer to protect water quality in neighboring lakes. Loss of wetlands in a watershed increases runoff into the lakes, reducing water quality. Wetlands are often vulnerable to development pressure, and many types of wetlands are hard to restore. To study a lake is to study its watershed, and in particular its associated wetlands or restorable wetlands. Water and nutrients flow through the system of lakes and wetlands, and plants and animals often need both lakes and wetlands to maintain healthy populations.

The size, shape, and location of an area are important. The larger a natural area is, the more likely it will sustain diverse populations of plants and animals. Many plants and animals are intolerant to disturbance and may require large undisturbed patches of habitat. Small, irregularly shaped natural area fragments may have more edge area than interior area, and these areas may be less important. However, those large areas with natural edges between habitat types provide important places for many plants and animals. Shorelines themselves are edges, as they represent the boundary between the lake habitat and the terrestrial habitat. Edges enable animals to access various habitats for nesting, foraging, or escape cover. Strategic conservation requires an integrated land- and lakescape approach that considers the influence of neighboring areas. Local changes and disturbances can have significant impacts on lake and river ecosystems. Think regionally in identifying sensitive lakeshore.

Shoreland and shorelines are often heterogeneous with critical habitat clustered. Shorelines may be comprised of a mix of windswept open areas, protected bays, and wetlands. Bays, because they are protected to some degree from wind and waves, often contain abundant vegetation. For example, they may contain a large portion of the valuable floating-leaf and emergent plant stands of a lake. Numerous fish species use these protected bays, wetland fringes, and the associated vegetative cover disproportionately to their availability. Fish prefer wetland embayments because they generally warm up faster in the spring, provide plentiful vegetative cover, and support high overall biological productivity. The green shallows are often used as spawning and nursery grounds. Some fish species, such as pugnose shiner, least darter, banded killifish, blacknose shiner, and blackchin shiner, are intolerant of disturbance and require large undisturbed patches of nearshore vegetation. Loons prefer to nest in these protected areas as well, and they come back to the same spots year after year. The spatial heterogeneity is also a result of human land and shoreline alteration. We urbanize and domesticate the landscape, change the drainage networks, and construct massive infrastructure. We eliminate and fragment native vegetation. Each of these human sources of lakeshore heterogeneity has environmental consequences. For a simple example, the fragmentation of vegetation often results in decreased nest success of some bird species, as birds forced to nest in habitat edges have higher predation from feral cats, squirrels, and crows (*Corvus brachyrhynchos*).

Connectivity is essential for ecosystems to function properly. The loss of connectivity through the addition of roads, buildings, and farms often fragments landscapes. Connectivity of natural areas allows animals to more easily move across the landscape, so animals with variable habitat requirements at different stages of their life cycle can find suitable places for feeding, mating, and rearing of young. Lake habitat quality is dependent on maintaining vegetated riparian zones and shorelines and connectivity to upland vegetation. Continual habitat loss breaks up landscapes of contiguous, diverse habitat for wildlife into increasingly smaller parcels and fragments. The fragments become isolated or disconnected, and the animals that depended on the diversity and connectivity suffer directly through habitat loss leading to population reduction or indirectly through the disruption of corridor habitats leading to increased mortality or decreased reproductive success. Habitat fragmentation becomes an issue in and of itself, and because the surroundings of the remaining fragments are becoming more intrusive and problematic, connectivity becomes more important in conservation. Connectivity is, we argue, also critical for sustainable planning and design of lakeshore living.

© ANDREA LEE LAMBRECHT

BALD EAGLES PERCHED IN A LARGE WHITE PINE AT THE SHORE.

PART TWO

Scientist, Writer, and Activist

Lakes are ecosystems defined by connectivity. Understanding ecosystems is necessary if we are interested in a rich lakeshore life. Some scientific insight helps. Despite considerable improvements in the field of science and mounting evidence to support scientific statements, too often we are unwilling to accept the derived facts. People invented the scientific method to think thoroughly about problems and to reliably understand the world around them. The strength of one's belief does not decide the facts. Science is a process to discover and revise facts about our world, even when our sensory capacity precludes discovery. We can't see, hear, or feel all forms of energy or structures of matter, but we create theories, equations, and machines to help us understand.

The scientific method and the statistical tools used in this process have evolved and expanded immensely in the last century. Whereas early scientific publications predominantly presented a collection of case studies where researchers used descriptive statistics, scientists generally recognize that strength of evidence also comes with the use of controls, randomization, replication, and statistical inference. "Control" refers to a group that is used for comparison when analyzing the results of an experiment or survey. For example, is the control group different from the group that received some sort of treatment? "Randomization" refers to selecting a random sample from a group and random assignment to a treatment. "Replication" refers to repeating tests or experiments or to having many samples. Testing requires replication—without replication one cannot be sure that the treatment caused the result. Perhaps it was merely coincidence that two groups were the same (or different) after treatment or after a period of time. "Statistical inference" involves modeling or making predictions based on data patterns and uncertainty. Today's scientific method has advanced to also include testing a suite of models to assess efficacy in predicting how the world works. Strength of evidence is central to science and the determination of reliable information.

Scientists in training learn the value of clear, focused, and concise writing. Addressing the ambiguous and deleting the extraneous are essential parts of scientific writing. Poetry is different, yet similar. Poetry is art that through choice of words attempts to tell a story or elicit an emotional response. Poetry employs ambiguity in the subtlest manner, creating complex spaces for interpretation. However, scientific writing and poems often share literary devices, such as conciseness, context, exposition, metaphor, parallelism, and special attention to word choice and language. In addition, the scientist and the poet both hope to produce works that are meaningful and useful to society. While there are not many scientist-poets, there are times when science is expressed quite poetically and poetry coincides with scientific tenets.

Aldo Leopold, Sigurd Olson, and William Whyte were scientists who blended science with poetry. They studied ecosystems and the human condition and were conservation activists addressing issues of their time. They were also environmental writers whose ideas continue to capture people's imagination. Their works provide insight and guidance on the fundamentals of rich lakeshore living. Aldo Leopold (1887–1948) created the field of wildlife management. His book *A Sand County Almanac and Sketches Here and There* gives the reader a view into his creative mind and into a conservation passion that provides a guiding light to all interested in the coexistence of people and nature. Sigurd Olson (1899–1982) was a science teacher, canoe country guide, ecologist, wilderness protection advocate, and Minnesota conservation writer. Through his writings and speeches, he became the poet of wilderness and an influential advocate for conservation. William Hollingsworth "Holly" Whyte Jr. (1917–1999) was a journalist, sociologist, and author of *The Organization Man.* Many of his writings focused on suburban development, urban revitalization, and human habitat, and he was influential in the promotion of tools to reduce urban sprawl.

The next three chapters walk through the lives of these thoughtful writers, and we gather, distill, and derive their philosophies and principles on lakeshore living as we have come to understand them today. Of course, their works are embedded in and indebted to a line of people who combined literary descriptions and experiences of nature with scientific analysis and advocacy. In North America, George Perkins Marsh, Henry David Thoreau, John Muir, and John Burroughs are prominent members of that list. Leopold and Olson were familiar with their writings and considered themselves part of the same lineage, of a budding movement. Henry David Thoreau, writing about his neighborhood lakes in "The Ponds" chapter in *Walden,* said: "A lake is the landscape's most beautiful and expressive feature. It is earth's eye; looking into which the beholder measures the depth of his own nature." Leopold and Olson likely concurred with this thought. A sense of sharing and of inescapable responsibility pervade their writing. Leopold came to consider the protection of our native landscapes as a moral imperative. Olson determined that it was in our spiritual best interest to have wilderness near and available. Whyte, a perceptive observer of human behavior in public spaces, came to believe that quality of life could only be found in communities that foster engagement with others and with nature.

The conservation movement continued to evolve into a broad environmental movement triggered by the publication of *Silent Spring* by Rachel Carson in 1962 and advanced by the report *The Limits to Growth* by Donella Meadows and others published in 1972. At the same time, urban planning was incorporating the thoughts of Whyte and others (e.g., Jane Jacobs, *The Death and Life of Great American Cities*). New Urbanism, conservation design, smart growth, and green urbanism all embraced aspects of the philosophy developed in the environmental writing tradition.

CHAPTER 3

Aldo Leopold and Living in Harmony with the Land

ALDO LEOPOLD WAS A NATURALIST, NATURAL RESOURCE MANAGER, PIONEERING ECOL-ogist, and land ethicist. His writings include *Report on a Game Survey of the North Central States* (1931), *Game Management* (1933), *A Sand County Almanac and Sketches Here and There* (1949), and *Round River: From the Journals of Aldo Leopold* (edited by his son, Luna Leopold, after Aldo Leopold's death). He published more than five hundred articles, and his work includes many essays and reports of some significance. Aldo Leopold was a restless and curious man who deliberately observed and documented nature, good and bad natural resource management, and ecological concepts. He expressed the need for human society to minimize its violence toward ecosystems, and he was a strong advocate for an ecological conservation conscience and ethic. Today, over sixty years after Aldo Leopold's death, natural resource managers are still inspired by the breadth and depth of his work. Aldo Leopold initiated and developed the field of wildlife management, was an important figure in the wilderness protection movement, educated landowners on ecological matters, and, in his later years, furthered a conservation ethic through a thoughtful reflection on his own career and life.

EXCURSIONS TO COUNTRYSIDE, MARSH LAKES, AND LAKE HURON

Aldo Leopold was born on January 11, 1887, to Carl and Clara Leopold in Burlington, Iowa. Carl was an avid hunter and the owner and manager of a company that made high-quality desks. Clara, Carl's first cousin, was an active mother who played piano, gardened, and figure skated. Clara was the daughter of Charles Starker, a Bavarian civil engineer who at Aldo's birth was Burlington's most prominent businessman, community supporter, and public servant. The Leopold family, later enlarged with a daughter and two more sons, lived a privileged life. Aldo's childhood home reflected the interests of his grandfather. His grandfather planted acres of red oak, spruce, and pines on the family estate overlooking the Mississippi River. The gardens harbored a wide variety of flowers. Carl's hunting and the family's weekend trips to the nearby countryside connected the family to the land, as did a milking cow, vegetable garden, orchard, and greenhouse.

Aldo Leopold grew up in the Midwest as it transitioned from frontier to a settled and developed region. The city of Burlington was situated on the banks of a still-untamed Mississippi

River, with its eight-mile-wide floodplain of marshes and ponds. At Aldo Leopold's birth, it was a modest city of twenty-five thousand people. Logs from upstream forests were still rafted in the river to the city's lumber company. The Iowa prairie, recently depopulated of bison, was still home to prairie chickens, and the bluff and upland oak and hickory groves held abundant wildlife. The railroad had recently become vital for the city, accelerating and amplifying the movement of farmers and goods.

Young Aldo Leopold, often with his younger brother, spent his time exploring the bluffs and boating the river. He was a good student and enjoyed the writings of Henry David Thoreau, Jack London, and Ernest Thompson Seton. At an early age, Aldo became interested in birds. His father, who hunted migrating waterfowl on the river and witnessed the decline of wildlife from the excesses of market hunting, developed his own conservation conscience and hunting ethics. Carl taught Aldo to hunt by the time he was a teenager, and they loved to roam the reed-covered backwaters of the Mississippi River.

The Mississippi River, prior to the construction of locks and dams, was a juxtaposition of braided channels, islands, backwater lakes, sloughs, and wetlands. The river landscape was diverse, with river terraces, old riverbanks, overbank deposits, sandbars, pools, riffles, and oxbow sloughs. The lush backwaters had water lilies, wild celery, arrowhead, smartweed, and numerous grasses, providing wildlife with food and habitat. In 1943 Aldo Leopold noted "the whole original ecology of the Mississippi was built on unstable water levels." As a boy he had observed ducks congregating in the backwater lakes in the spring and fall, feeding on seeds and tubers from vegetation that benefited from low water levels. Many of these backwater lakes are gone now, destroyed by the completion of Lock and Dam 19 in 1913 and the construction of numerous levees. Late in his life when he returned to these areas, Leopold said: "The job was so complete that I could not even trace the outlines of my beloved lakes and sloughs under their new blanket of cornstalks. I like corn, but not that much. Perhaps no one but a hunter can understand how intense an affection a boy can feel for a piece of marsh. My home town thought the community enriched by the change, I thought it impoverished."

Carl and Clara Leopold regularly took the family on a six-week vacation to Marquette Island at the northwest end of Lake Huron. This lakeshore living provided Aldo opportunities to explore the north woods of hemlock, balsam fir, cedar, and white pine and the lake and island with its points, bays, and inlets. These vacations likely sparked Aldo's interest in wildness and fish. Aldo enjoyed fishing for yellow perch, smallmouth bass, northern pike, and muskellunge (*Esox masquinongy*).

Marquette Island is the largest island of the Les Cheneaux archipelago, which consists of thirty-six islands along twelve miles in Lake Huron, thirty miles northeast from the Straits of Mackinac. The island is still a popular recreation area, and the channel is dotted with cottages and boathouses, many of which date to the early 1900s. The archipelago forms many bays and harbors. Aldo made maps of Marquette Island, showing bays and trails and illustrating them with the trees and the birds he observed. Concerned local citizens, the Little Traverse Conservancy, and the Nature Conservancy recently protected part of this island as the Aldo Leopold Nature Preserve.

Aldo Leopold's high school in Burlington provided him with a good education in the natural sciences. His bird studies and natural history writings during this time became more scientific and included details of bird behavior and bird songs. He soon became interested in the new science of forestry management. When Aldo was sixteen years old, the family traveled to Colorado and Montana and visited Yellowstone; as he often did, Aldo documented the trip in a diary. Shortly after returning to Burlington, Aldo left for a year at a New Jersey college preparation boarding school. There he conducted extensive excursions into the countryside, noting and recording geology, natural features, trees and other plants, and the birds he observed. He wrote profusely to his parents, describing his outdoor exploits and experiences, writing in a style of optimism, grace, and wit that would remain recognizable for his entire life.

After another summer at Burlington and Marquette Island, Leopold began college at Yale University, the only university at the time offering coursework and training in forestry management. During his freshman year, he continued his countryside excursions and letters to home, but he spent more time with studies, school activities, and social events. His sophomore year included studies of land surveying and plant taxonomy. Before his third year at Yale, Aldo and his father spent time hiking the hills and valleys of New England, including Vermont's Green Mountains, fishing the rivers, and camping. The following summer found Aldo at the Yale Forest School summer camp in Pennsylvania. There he practiced surveying, mapping, and forest thinning and took pleasure in hiking and fishing. Further studies on silviculture and botany marked his senior year. He was also immersed in a variety of extracurricular activities, such as work, debate, club functions, and visiting friends. Following graduation, he returned home and once again went with the family to Lake Huron.

Aldo entered Yale Forest School as a master's student in 1908. In 1909, after being schooled in timber measurement and forest regulation, he began work for the U.S. Forest Service. The Forest Service was led by Gifford Pinchot, the agency's first chief and ardent supporter of the Yale Forest School. Aldo took his first professional position at the Apache National Forest, a wilderness forest in the Arizona Territory. After two years, Aldo was promoted to a supervisory position at the Carson National Forest north of Santa Fe. This forest was heavily used and overgrazed by cattle, and Aldo witnessed the consequences of poor land management on wildlife, plants, and water.

Aldo Leopold's formative years benefited from multiple experiences in diverse landscapes, formal training, thoughtful and caring parents, and exposure to many kinds of people. The lakeshore experiences of his youth and his work in the southwest ingrained in him the knowledge that the health of the countryside and its people is a direct function of the health of the watershed.

MADISON LAKES

In 1924, after working on erosion, fish and wildlife management, wilderness protection, and watershed management for over a decade in the southwest, Aldo Leopold, married with four children and thirty-seven years old, a respected leader and innovator for the U.S. Forest Service, moved his family to Madison, Wisconsin. A year earlier, Aldo had produced a watershed handbook. Forest Service field staff used his handbook to identify and correct erosion and watershed problems in the southwest. In the year of Leopold's move to Madison, a scientific journal published his article summarizing his ecological understanding of the Arizona brushlands and a magazine published his essay on the perilous effects of poor land use on erosion and sedimentation of rivers, flowages, and reservoirs. In the latter, Aldo Leopold wrote: "The day will come when ownership of land will carry with it the obligation to so use and protect it with respect to erosion that it is not a menace to other landowners and the public . . . it is cheaper to prevent erosion than to cure it, and the cost of such prevention must some day be passed on uniformly by all landowners to all consumers of their products. But enforced responsibility of landowners is of the future."

Aldo advocated more vegetation along watercourses. He also noted the tragedy of the commons on open access lands, forty-four years before Garrett Hardin's classical work on the issue. The tragedy of the commons occurs in both space and time; a landowner who exploits the land and degrades the hydrology negatively affects his neighbors both now and in the future.

Before beginning his new job at the U.S. Forest Products Laboratory, Aldo Leopold, along with his two brothers and eldest son, traveled to northern Minnesota to canoe the Quetico-Superior lakes along the U.S.-Canada border. He would return to these lakes again. He kept a journal during the fourteen-day excursion, describing the beauty of the loon's call, the fish he caught, and many other experiences. In the fall, the family hunted ducks and other waterbirds in the wetlands, ponds, and lakes of Dane County. They soon became avid archers.

The Madison lakes include the four lakes of the Yahara River: Mendota, Monona, Waubesa, and Kegonsa. Downtown Madison is on the isthmus between Mendota and Monona Lakes, and the University of Wisconsin–Madison stretches along the south shore of Lake Mendota. The lakes were formed when a glacier covered the Yahara valley, depositing rock and sediment that blocked the stream valley. Edward A. Birge and Chancey Juday were leaders in the development of the field of limnology, and they founded a school of limnology at the University of Wisconsin–Madison in 1909, where a tradition of limnology excellence continues today. Birge and Juday conducted much of their pioneering work on Lake Mendota beginning in 1908.

Lake Mendota drains a large watershed of over two hundred square miles. The early land use was mostly agricultural and is now increasingly urban. Today Lake Mendota experiences serious blue-green algae blooms due to high phosphorus inputs from past and current agricultural practices and urban runoff in the watershed. Common carp and Eurasian water milfoil have found their way into the lake. Loss of native aquatic plants has occurred, caused by increasing eutrophication and carp populations. The lake has a diminished diversity of fish; a third of the original fish species have been eliminated from the lake. Recently, community leaders across the watershed have developed an implementation plan to achieve a 50 percent reduction in phosphorus runoff to the lake.

Once in Madison, Aldo Leopold did not take long to get deeply involved with local and regional conservation issues. He was active in the recently created Izaak Walton League, and he worked on issues related to wildlife management, reducing wood waste, and wilderness protection. He had become a nationally recognized advocate for wilderness protection and spoke on the subject at important events. Aldo wrote a far-reaching essay for the applied scientific *Journal of Land and Public Utility Economics* (now called *Land Economics*). In this paper, titled "Wilderness as a Form of Land Use" (1925), he argued for a comprehensive policy of wilderness protection. Leopold stated:

> The first idea is that wilderness is a resource, not only in the physical sense of raw materials it contains, but also in the sense of a distinctive environment which may, if rightly used, yield certain social values. . . . Fourth, wilderness exists in all degrees, from the little accidental wild spot at the head of a ravine in a Corn Belt woodlot to vast expanses of virgin country—"where nameless men by nameless rivers wander and in strange valleys die strange deaths alone." What degree of wilderness, then, are we discussing? The answer is, *all degrees*. Wilderness is a relative condition. As a form

of land use it cannot be a rigid entity of unchanging content, exclusive of all other forms. On the contrary, it must be a flexible thing, accommodating itself to other forms and blending with them in that highly localized give-and-take scheme of land-planning which employs the criterion of "highest use." By skilfully adjusting one use to another, the land planner builds a balanced whole without undue sacrifice of any function, and thus attains a maximum net utility of land.

Aldo Leopold became passionately involved in the protection efforts for Minnesota's canoe country within the Superior National Forest, a place that had touched him during his summer excursions. Ernest Oberholtzer and Sigurd Olson were involved in the same fight.

Aldo and his wife, Estella, had their fifth child in 1927 (eldest to youngest: [Aldo] Starker, Luna, Nina, [Aldo] Carl, and Estella). While happy with his home life, Leopold was interested in a career change, preferably to a job in wildlife management. In 1928, he left his job with the Forest Products Lab to spend four years under contract with a gun and ammunition consortium. Here he conducted extensive game animal surveys to identify major trends in populations and habitats, stressors on these animals, and management and governance of game in the Midwest states. He compiled information on animal ranges, habitat, population cycles, natural resource management, population dynamics, and game animal policy. He talked to farmers, natural resource management agency staff, hunters, university professors, and conservationists. During the Great Depression, Leopold shepherded the passage of a national game animal policy, wrote the first ever textbook on game animal management, and broadened the discourse on wilderness protection through lectures and commentaries. In response to criticism about the national game animal policy, Aldo wrote in the scientific journal *The Condor* (1932): "Mine is a system of *proposed public actions* designed to fit the unpleasant fact that America consists largely of business men, farmers, and 'Rotarians,' busily playing the national game of economic expansion. Most of them admit that birds, trees, and flowers are nice to have around, but few of them would admit that the present 'depression' in waterfowl is more important than the one in banks, or that the status of the blue goose has more bearing on the cultural future of America than the price of U.S. Steel."

This national policy on wildlife management advocated the use of the economic system to address landowners who restricted access to their lands. The framework included payment for hunting access to private lands in areas where public land was too expensive to purchase. In his response, Aldo continued:

When I submit these thoughts to a printing press, I am helping cut down the woods. When I pour cream in my coffee, I am helping drain a marsh for cows to graze, and to exterminate the birds of Brazil. When I go birding or hunting in my Ford, I am devastating an oil field, and re-electing an imperialist to get me rubber. Nay more: when I father more than two children I am creating an insatiable need for more printing presses, more cows, more coffee, more oil, and more rubber, to supply which more birds, more trees, and more flowers will either be killed, or what is just as destructive, evicted from their several environments.

What to do? I see only two courses open to the likes of us. One is to go live on locusts in the wilderness, if there is any wilderness left. The other is surreptitiously to set up within the economic Juggernaut certain new cogs and wheels whereby the residual love of nature, inherent even in "Rotarians," may be made to recreate at least a fraction of those values which their love of "progress" is destroying. A briefer way to put it is: if we want Mr. Babbitt to rebuild outdoor America, we must let him use the same tools wherewith he destroyed it. He knows no other.

Mr. Babbitt is the main character in Sinclair Lewis's 1922 fictional, satirical book *Babbitt.* Though a successful realtor, he is lost at midlife in his materialism and mindless societal conformity. Later in life, Leopold came to recognize some of the shortcomings of economic tools such as incentives, land-abuse fines, and public subsidies, and he began to include, and then emphasize, individual and community responsibility for conservation.

Aldo Leopold broadened these conservation thoughts, dependent on moral values, with his remarkable paper "The Conservation Ethic." Published in the *Journal of Forestry* in 1933, this paper was Aldo's early, yet refined, attempt to share his thoughts on the changes needed in our society to live within our means and to save shards of nature for human benefit. Aldo Leopold wrote:

> There is as yet no ethic dealing with man's relationship to land and to the non-human animals and plants which grow upon it. Land, like Odysseus' slave-girls, is still property. The land-relation is still strictly economic, entailing privileges but not obligations.
>
> The extension of ethics to this third element in human environment is, if we read evolution correctly, an ecological possibility. It is the third step in a sequence. The first two have already been taken. Civilized man exhibits in his own mind evidence that the third is needed. For example, his sense of right and wrong may be aroused quite as strongly by the desecration of a nearby woodlot as by a famine in China, a near-pogrom in Germany, or the murder of the slave-girls in ancient Greece. Individual thinkers since the days of Ezekiel and Isaiah have asserted that the despoliation of land is not only inexpedient but wrong. Society, however, has not yet affirmed their belief. I regard the present conservation movement as the embryo of such an affirmation. I here discuss why this is, or should be, so.

Fifteen years later Leopold revised this essay to become the enduring "The Land Ethic" in *A Sand County Almanac.*

Aldo Leopold joined the faculty at the University of Wisconsin–Madison in 1933 as a professor of wildlife game management. In his earlier jobs, he led, guided, and administered large organizations. Now Leopold was primarily a teacher to young students and Wisconsin farmers and landowners. He was engaged in committees, collaborative projects, and conservation organizations. One such assignment was for a Presidential Committee on Wild Life Restoration, where Leopold presented information to President Franklin D. Roosevelt. Leopold wrote about the New Deal with regard to natural resources management in "Conservation Economics," published in 1934 in the *Journal of Forestry.* Leopold's primary point was that overreliance on acquiring public lands for conservation would not solve the problems of game bird supply or land erosion at the scale necessary. He stated that only buying land for the public was "as effective as buying half of the umbrella." He noted that "conservation use of every acre on every watershed in America" was needed "to assure the physical integrity of America." He also saw the benefit of using public dollars to fund good land use; this belief is known today as payment for ecosystem services. Leopold concluded: "This paper forecasts that conservation will ultimately boil down to rewarding the private landowner who conserves the public interest. It asserts the new premise that if he fails to do so, his neighbors must ultimately pay the bill."

Leopold worked to restore the lands for the University's Arboretum with horticulturalist and landscape architect G. William Longenecker and botanist Norman Fassett. In 1934, Leopold helped found the Wilderness Society. In 1935, the Leopold family bought an eighty-acre abandoned alluvial farm on the Wisconsin River in Sauk County. The land was at the southern end of the central Wisconsin sand counties. The only structure there at the time was a chicken coop, later reclaimed by the family to become "the shack." The first visitors to the farm included Paul Errington (Aldo's colleague from Iowa State University) and his wife, and two of Errington's students, Frederick and Frances Hamerstrom. The shack was the family's weekend cabin on the river shores. It also served as headquarters for their restoration and forestation of the surrounding land, a difficult enterprise during the dry Dust Bowl years of the 1930s. Today, his family place and surrounding lands are protected by a land trust. Neighbors and other citizens formed the Leopold Memorial Reserve, which consists of over 1,900 acres of private land. The lands were included in an Important Bird Area, which is appropriate given Leopold's love of birds. The shack is on the National Register of Historic Places, the only chicken coop on the register. The shack is less than thirty miles from the childhood home of John Muir, conservationist and founder of the Sierra Club, on Ennis Lake. These lands are now a 150-acre county park and state natural area used for nature study and fishing.

Aldo Leopold used an opportunity at a University of Wisconsin event in 1939 to illuminate his views on the meaning of conservation; the speech was later published in *American Forests*. Aldo delivered the following words:

> Conservation means harmony between men and land. When land does well for its owner, and the owner does well by his land; when both end up better by reason of their partnership, we have conservation. When one or the other grows poorer, we do not.
>
> Few acres in North America have escaped impoverishment through human use. If someone were to map the continent for gains and losses in soil fertility, waterflow, flora, fauna, it would be difficult to find spots where less than three of these four basic resources have retrograded; easy to find spots where all four are poorer than when we took them over from the Indians. . . . When [Dr. Charles] Van Hise said "Conservation is wise use," he meant, I think, restrained use.
>
> Certainly conservation means restraint, but there is something else that needs to be said. It seems to me that many land resources, when they are used, get out of order and disappear or deteriorate before anyone has a chance to exhaust them. . . . Consider the growing dependence of fishing waters on artificial restocking. A big part of this loss of toughness inheres in the disordering of water by erosion and pollution.
>
> Conservation, then, is keeping the resource in working order, as well as preventing over-use. Resources may get out of order before they are exhausted, sometimes while they are still abundant. Conservation, therefore, is a positive exercise of skill and insight, not merely a negative exercise of abstinence or caution.

Leopold went on to say that the landscape "is the owner's portrait of himself" and working lands should be "a mixture of wild and tame attributes, all built on a foundation of good health."

At the university, Aldo Leopold was pioneering the scientific field of ecology. Defining the scope of ecology in conservation was Leopold's passion at this point in his career. The ray of light Aldo was following illuminated solutions to erosion, wildlife management, and, more comprehensively, total land health. Leopold, in a plenary address to the Society of American Foresters and the Ecological Society of America in Milwaukee, Wisconsin, in 1939 (published as "A Biotic View of the Land" by the *Journal of Forestry*), said:

Ecology is a new fusion point for all the natural sciences. It has been built up partly by ecologists, but partly also by the collective efforts of the men charged with the economic evaluation of species. The emergence of ecology has placed the economic biologist in a peculiar dilemma: with one hand he points out the accumulated findings of his search for utility, or the lack of utility, in this or that species; with the other he lifts the veil from a biota so complex, so conditioned by interwoven cooperations and competitions, that no man can say where utility begins or ends. No species can be "rated" without the tongue in cheek; the old categories of "useful" and "harmful" have validity only as conditioned by time, place, and circumstance. The only sure conclusion is that the biota as a whole is useful, and biota includes not only plants and animals, but soils and waters as well.

Leopold, the teacher, then helps the audience with some mental models of ecosystems:

> To the ecological mind, balance of nature has merits and also defects. Its merits are that it conceives of a collective total, that it imputes some utility to all species, and that it implies oscillations when balance is disturbed. Its defects are that there is only one point at which balance occurs, and that balance is normally static.
>
> If we must use a mental image for land instead of thinking about it directly, why not employ the image commonly used in ecology, namely the biotic pyramid? . . . The upward flow of energy depends on the complex structure of the plant and animal community, much as the upward flow of sap in a tree depends on its complex cellular organization. Without this complexity normal circulation would not occur. Structure means the characteristic number, as well as the characteristic kinds and functions of the species.

Through the 1930s and 1940s, Aldo Leopold took on many graduate students (twenty-seven total). They included Franklin Schmidt (prairie chicken research on the central sand counties), Arthur Hawkins (quail research in southern Wisconsin), Hans Albert Hochbaum (canvasback duck research at the Delta Marsh on Lake Manitoba), Frederick and Frances Hamerstrom (prairie chicken research on the central sand counties), Joseph Hickey (whom Leopold later invited to become the second professor in UW–Madison's Department of Wildlife Management), Irven Buss (upland sandpipers and pheasants), and Robert McCabe (pheasants). War rumors permeated society, and fear of the future was high when Germany invaded Poland on September 1, 1939. Acutely aware of pressing world events, Leopold was more reflective and poetic. He wrote short nature stories, including ones about the southwest. His professional writings sought to explain large ecological concepts. They occasionally included thoughts on politics and governance, but more often incorporated ideas on potential solutions to many ecological problems of the period (many of which remain intractable today). Trips to the shack with his family gave Aldo Leopold much-needed rest from a schedule filled with committees, organizational assignments, correspondence, and teaching.

In 1940, Aldo Leopold presented a talk entitled "Lakes in Relation to Terrestrial Life Patterns." Leopold stressed the fact that the land and the lake were connected:

> Soil and water are not two organic systems, but one. Both are organs of a single landscape; a derangement in either affects the health of both. . . . All land represents a downhill flow of nutrients from the hills to the sea. This flow has a rolling motion. Plants and animals suck nutrients out of the soil and air and pump them upward through the food chains, the gravity of death spills them back

into the soil and air. Mineral nutrients, between their successive trips through this circuit, tend to be washed downhill. Lakes retard this downhill wash, and so do soils.

Leopold explained that humans often shorten food chains. In a nonanthropogenic chain, a nutrient may be bound in a bur oak for a long period of time before being passed to a squirrel, then to a red-tailed hawk, and then to other organisms. In the human-domesticated landscape it may take only a year for the nutrient to pass from corn to human to sewer to lake.

Aldo Leopold questioned why lakeshore citizens allowed "algae control, swimmer's itch control, and the planting of any fish able to swim" in their lakes. He noted that these and other activities were "hasty tinkerings." Leopold concluded his talk, reiterating his opening ideas, with "soil health and water health are not two problems, but one." It was common then to think of these two systems (land and lake) as two different ecosystems. Research into groundwater dynamics, energy flow, and nutrient cycles has shown that the land and the water are best viewed as one system. To solve many lake problems one must look landward first, and Leopold understood this.

The United States entered World War II late in 1941, changing many people's lives. Aldo Leopold's son Carl enlisted in the Marines, deployed in the south Pacific, and son Luna enlisted in the Army. Many of Leopold's students went to war. Trips to the shack were reduced due to gasoline rationing. Leopold would spend some of his time on a collection of essays for a book he was developing.

The end of World War II brought a surge of students and little time for his essay collection project. Professor Leopold called on former students to assist with the heavy teaching load. Deep respect from his peers allowed Aldo Leopold to speak candidly about conservation problems and solutions. On June 27, 1947, in Minneapolis, Minnesota, Aldo Leopold delivered a concise, powerful speech to members of the Garden Club of America on the need for the individual to cultivate an ecological conscience (the speech was published as "The Ecological Conscience"). In his talk, Leopold defined ecology as the science of communities and ecological conscience as the ethics of community life. Leopold noted the slow spread of conservation, and he asserted the following:

> The basic defect is this: we have not asked the citizen to assume any real responsibility. We have told him that if he will vote right, obey the law, join some organizations, and practice what conservation is profitable on his own land, that everything will be lovely; the government will do the rest.
>
> This formula is too easy to accomplish anything worthwhile. It calls for no effort or sacrifice; no change in our philosophy of values. It entails little that any decent and intelligent person would not have done, of his own accord, under the late but not lamented Babbitian code.
>
> No important change in human conduct is ever accomplished without an internal change in our intellectual emphases, our loyalties, our affections, and our convictions.

Leopold stated that people must have open minds. When confronted with new information and facts, one must change one's opinions and adapt to the new realities, and have "a capacity to study and learn, as well as to emote about the problems of conservation." On the evolution of an ecological conscience, he concluded by asserting: "In such matters he should not worry too much about anything except the direction in which we travel. The direction is clear, and the first step is to throw your weight around on matters of right and wrong in land-use. Cease being intimidated by the argument that a right action is impossible because it does not yield maximum profits, or that a wrong action is to be condoned because it pays. That philosophy is dead in human relations, and its funeral in land-relations is overdue."

Suffering from a recurring medical ailment that forced him to spend less time on his many professional commitments, Leopold was able to spend more time at the shack. Here he sharpened his literary craftsmanship to merge a broad range of conservation essays and writings into a book. On April 14, 1948, Oxford University Press notified Aldo Leopold that they were interested in publishing his book. That same afternoon Aldo, his wife, and youngest daughter Estella headed to the shack for several days of tree planting (white and red pines). On April 21, a grass fire spread from a neighbor's trash fire and began heading toward Leopold's past plantings. In the act of suppressing the fire, Aldo Leopold was struck with a heart attack. In the commotion and confusion of the fire, no one noticed Aldo's condition. Aldo Leopold died at the scene of the fire.

After his death, his friends, comprised of his closest students and family, worked together on the book manuscript Aldo had left near completion. John Hickey worked with Oxford University Press. Luna Leopold was the chief editor. Estella Leopold, the Hamerstroms, and others reviewed the essays and advanced minor editorial changes. *A Sand County Almanac and Sketches Here and There* was published in 1949 and still stands as a classic in the conservation field.

LAKESHORE LIVING WITH AN ECOLOGICAL CONSCIENCE

Leopold's work and writings provide a foundation for an ethical, full, rich lakeshore life. This way of lakeshore living strives to be harmonious with the environment and minimizes the derangement of land and lake. Real practical wisdom takes time to accumulate, and some of us do a better job of synthesizing our experiences and turning this information into advice. Aldo Leopold's wisdom is better understood with repeated readings. Science has advanced since Leopold's days, but many principles he revealed are still valid today. Leopold provides the first-order principle for a full, rich lakeshore life in his summary precept in *A Sand County Almanac:* "A thing is right when it tends to preserve the integrity, stability, and beauty of the biotic community. It is wrong when it tends otherwise." He provides four second-order principles for conserving nature, which are listed below. These four originate from a 1946 handwritten manuscript by Aldo Leopold, first published in 1999 as "The Land Health Concept and Conservation" in *For the Health of the Land.* Lastly, we derived a fifth principle from his works.

1. Cease Throwing Away Its Parts.

    "Communities are like clocks, they tick best while possessed of all the cogs and wheels." ["Last Stand," *The River of the Mother of God*]
    "If the land mechanism as a whole is good, then every part is good, whether we understand it or not. If the biota, in the course of aeons, has built something we like but do not understand, then who but a fool would discard seemingly useless parts? To keep every cog and wheel is the first precaution of intelligent tinkering." ["The Round River," *A Sand County Almanac*]
    "Everybody knows, for example, that the autumn landscape in the north woods is the land, plus a red maple, plus a ruffed grouse. In terms of conventional physics, the grouse represents only

a millionth of either the mass or the energy of an acre. Yet subtract the grouse and the whole thing is dead." ["Guacamaja," *A Sand County Almanac*]

"Darwin gave us the first glimpse of the origin of the species. We know now what was unknown to all the preceding caravan of generations: that men are only fellow-voyagers with other creatures in the odyssey of evolution. This new knowledge should have given us, by this time, a sense of kinship with fellow-creatures; a wish to live and let live, a sense of wonder over the magnitude and duration of the biotic enterprise." ["On a Monument to the Pigeon," *A Sand County Almanac*]

2. Handle It Gently.

"This leads to the 'rule of thumb' which is the basic premise of ecological conservation: the land should retain as much of its original membership as is compatible with human land-use. The land must of course be modified, but it should be modified as gently and as little as possible." ["Conservation: In Whole or in Parts," *The River of the Mother of God*]

"My guess here is that the less violent these [land] conversions, the more likely they are to be durable, and the less likely they are to exhibit unforeseen repercussions." ["The Land Health Concept and Conservation," *For the Health of the Land*]

"Land is unequally sensitive. . . . Lands differ in their toughness." ["Biotic Land-Use," *For the Health of the Land*]

"Biotas seem to differ in their capacity to sustain violent conversion." ["The Land Ethic," *A Sand County Almanac*]

"A land ethic of course cannot prevent the alteration, management, and use of these 'resources,' but it does affirm their right to continued existence, and, at least in spots, their continued existence in a natural state." ["The Land Ethic," *A Sand County Almanac*]

3. Recognize that Its Importance Transcends Economics.

"'A refined taste in natural objects' perceives that the economic issue is a separate consideration." ["The Round River," *A Sand County Almanac*]

"Quit thinking about decent land-use as solely an economic problem. Examine each question in terms of what is ethically and aesthetically right, as well as what is economically expedient. . . . It of course goes without saying that economic feasibility limits the tether of what can or cannot be done for land. It always has and it always will. The fallacy . . . is the belief that economics determines all land use. This is simply not true. An innumerable host of actions and attitudes, comprising perhaps the bulk of all land relations, is determined by the land-users' tastes and predilections, rather than by his purse." ["The Land Ethic," *A Sand County Almanac*]

4. Don't Let Too Many People Tinker with It.

"The combined evidence of history and ecology seems to support one general deduction: the less violent the man-made changes, the greater the probability of successful readjustment in the [biotic] pyramid. Violence, in turn, would seem to vary with human population density; a dense population requires a more violent conversion of land." ["The Land Ethic," *A Sand County Almanac*]

"It is unthinkable that we shall stabilize our land without a corresponding stabilization of our density." ["The Land Health Concept and Conservation," *For the Health of the Land*]

"Why not seek for quality in place of ciphers in human populations?" ["Ecology and Politics," *The River of the Mother of God*]

5. Retain and Restore Nature's Health—Stable Soil, Natural Hydrology, and Habitat.

"The [biotic] pyramid is a tangle of chains so complex as to seem disorderly, yet the stability of the system proves it to be a highly organized structure. Its functioning depends on the co-operation and competition of its diverse parts." ["The Land Ethic," *A Sand County Almanac*]

"It is generally understood that when soil loses fertility, or washes away faster than it forms, and when water systems exhibit abnormal floods and shortages, the land is sick." ["Wilderness," *A Sand County Almanac*]

"In short, we face not only an unfavorable balance between loss and gain in habitat, but an accelerating disorganization of those unknown controls which stabilize the flora and fauna, and which, in conjunction with stable soil and a normal regimen of water, constitute land-health." ["The Outlook for Farm Wildlife," *The River of the Mother of God* and *For the Health of the Land*]

CHAPTER 4

Sigurd Olson and Protecting Wilderness

SIGURD OLSON WAS A MAN OF THE LAKES. ALL HIS LIFE, HE WAS INSPIRED, EVEN obsessed, with lake wilderness, and he often reflected on this obsession. Familiar characters in his narrative range from the French-Canadian voyageurs (he took pride in the honorific title "bourgeois" [leader] given to him by his friends) to the animals, trees, rocks, lakes, rivers, and portages. He appreciated the whole network of wet places that made the wilderness accessible for voyageurs and modern visitors alike. Sigurd Olson was very attached to the canoe country, presenting it as the last wilderness before the far West or high North.

Sigurd Olson and his elder colleague Ernest Oberholtzer were deeply involved in the protection of wilderness. Although their strategies differed, they agreed that the real challenge was found in the management of areas for both humans and nature. Whereas Oberholtzer was a landscape architect by education, Olson studied geology and ecology, and the difference can easily be discerned in their writings and activities. Both shared a passion for the wilderness, for exploration, for the labyrinthine waterscapes of the Quetico-Superior area, straddling the Canadian-U.S. boundary. Both worked tirelessly for the preservation of the lake wilderness, first under the leadership of Oberholtzer (prior to 1947) and later under Olson's guidance. Although they did not use the terms "mixed-use" or "multiple land use," they strove to find ways to meet the needs of both humans and the environment.

Oberholtzer utilized plans and design in his lobby work, as part of his strategy to convince people that more was possible, that an area could be developed in many different ways, and that many activities could be accommodated without jeopardizing the character of the landscape. (Unfortunately, most of these plans have been lost or are hidden in yet-unexplored archives.) Sigurd Olson took a different approach, offering advice primarily through his writings.

Sigurd Olson was a late bloomer as a writer. He wrote his first book, *The Singing Wilderness* (1956), when he was in his late fifties, but his poetic sensibility was evident much earlier. He recorded in diaries and in early articles that both sweeping vistas and minute details in nature could capture his attention, touch him emotionally, and bring him to an understanding that these elements are all part of a larger whole. About returning to a favorite lake Olson wrote: "Then at last I was back, and as I paddled along and saw the old familiar reaches of blue, the islands riding at anchor in the distance, the gnarled old trees and lichen-covered cliffs, it seemed as though I had never been away." This appreciation for all levels of nature came as a slow revelation, according to *Open Horizons,* largely an autobiography. Olson's reflection on the construction of his cabin and the landscaping of its surroundings expresses an almost

51

guilty awareness of his impact on the landscape. Still, he built the cabin, reusing much of an old Finnish structure used as a chicken coop at the time. Olson's second book, *Listening Point,* offers, perhaps more than any of his other books, practical implications of his ideas. The poetic quality of *The Singing Wilderness* was arguably never surpassed by Olson again. His later works like *Reflections from the North Country* and *Of Time and Place* might not always maintain the same level of prose writing, but the ideas are mature, deep, and meaningful. Olson's philosophical reflections address real-world issues. These issues should be confronted and resolved by planners, landscape architects, and lake lovers.

GETTING TO CANOE COUNTRY

Sigurd Olson was born on April 4, 1899, in Chicago to Ida May and Lawrence J. Olson. Seeking freedom to practice his Baptist faith and work, Lawrence (Lars Jakob Olsson) had emigrated from the lake and mountain province of Dalarna, Sweden, at the age of nineteen. Lawrence headed to Alexandria, Minnesota, where a relative lived, and he worked as a farm hand and carpenter for a year before entering Chicago's Morgan Park Seminary. Upon graduation, Lawrence became the fundamentalist minister of the Swedish Baptist Church of St. Cloud, Minnesota. Ida May Cederholm, a Swedish immigrant living in Brainerd, Minnesota, met Lawrence Olson when he was visiting the Brainerd lakes area, and the two were married shortly thereafter. The couple had three sons when they moved in November 1906 to Sister Bay on Wisconsin's Door County peninsula, which extends into Lake Michigan.

Sigurd Olson grew up around the lakeshores of Wisconsin. He was seven years old when his family left Chicago, and it was at Lake Michigan that he would later say he "heard the singing wilderness for the first time." For three years the family lived near Sister Bay, where Sigurd would take excursions into the wild areas near their rural two-story home. Sigurd learned to fish and to appreciate the sound of gulls, the taste and smell of fresh apples, and the sight of waves touching the shore. Later, the Olson family moved to northern Wisconsin, living first in Prentice, then in Ashland on the shore of Lake Superior. In Prentice, Sigurd spent his days fishing and hunting. He fished rock bass along the Jump River and brook trout in a local cold-water creek. He hunted grouse and snowshoe hares. In Ashland, Sigurd explored, hunted, and came to love Fish Creek Slough, which harbored ducks and swans, and he visited the Apostle Islands on Lake Superior.

Sigurd attended Northland College in Ashland from 1916 to 1918, majoring in agriculture. He did well in school and took to the idea of farming as a profession. During the summers of 1917 and 1918, Sigurd worked for Soren Uhrenholdt at his four-hundred-acre dairy and potato farm along the Namekagon River in northern Wisconsin. Uhrenholdt, a Danish immigrant, was a highly respected, progressive farmer who practiced forest conservation. Sigurd worked to clear land for the plow, using an ax and a team of horses. He fished and hunted with Soren's neighbors and son, and he became good friends with Soren and his daughter Elizabeth. Sigurd was an active student during his second year at Northland College. He was involved with the campus YMCA, choir, basketball team, newspaper, and the drama club. After his second year at Northland College, Sigurd began work at the TNT factory in Ashland, which was the world's largest TNT producer during World War I.

In the fall of 1918 Sigurd attended the University of Wisconsin–Madison to finish his agricultural degree. He also enlisted into an Army training program that provided combat training to college men. His training was short-lived as the war ended in November and he was discharged in December. In Madison, overlooking Lake Mendota, Sigurd confronted his fundamentalist Baptist religion and concluded that he had not yet developed a personal sense of purpose, a condition that plagued Sigurd for many years and often left him melancholy. Sigurd took a summer job working for the Wisconsin Geological Survey, which triggered greater interest and studies in geology at the university. He graduated in 1920 with a degree in agriculture (animal husbandry), with average grades and little passion for the science of farming. He was offered a job teaching high school classes on agriculture and geology for a school district on the iron range of northern Minnesota. Sigurd accepted the offer as it provided an opportunity for exploration of lakes and woods. He took the train from Madison north to Nashwauk, where he found living quarters and began his teaching career. Here, Sigurd lived for the weekends. After Friday's classes were dismissed, Sigurd left for the woods and lakes, camping with minimum supplies to allow him to hunt and explore as he wished. His long-distance relationship with Elizabeth Uhrenholdt grew, and they planned to marry.

Friends and neighbors encouraged Sigurd to head northeast from Nashwauk to the waterway wilderness along the Ontario-Minnesota border. When the school year ended in June 1921, he and some friends rented canoes and headed west past the forest clear-cutting to Basswood, Knife, Sea Gull, Saganaga, and Sawbill Lakes. Sigurd Olson fell in love with this wilderness—the granite outcrops, diverse shorelines, and thick woods. The canoe trip was cut short as Sigurd met Elizabeth at the Uhrenholdt farm for an August wedding. The couple honeymooned in the canoe country that Sigurd had just discovered. Many years later, in 1947, part of the Uhrenholdt farm was dedicated as the Uhrenholdt Memorial State Forest to protect an old growth stand of white pine and other large trees in memory of the Uhrenholdt family that conserved this place.

During his first year of marriage, Sigurd considered careers besides teaching, such as being a writer or a geologist. Sigurd was a competent teacher, but he felt trapped indoors. Sigurd enrolled back at the University of Wisconsin–Madison with a geology major, while Elizabeth taught at an elementary school in Hayward, Wisconsin. Sigurd continued to take basic teaching training courses, but geology was his new passion. Charles Kenneth Leith and Alexander Winchell taught Sigurd's geology courses. Both men were highly respected scientists. Leith had surveyed the Precambrian formation of the iron range of Minnesota and shared Sigurd's love of the Ontario-Minnesota canoe country. Winchell was a noted mineralogist. Leith and Winchell shaped how Sigurd would view the world by providing a broad picture of the earth with its long time scales and large events. Despite Sigurd's newfound career contentment, he left the university after his first graduate semester to find a job to support his family. Fortuitously, the superintendent of Ely, Minnesota, was in Madison looking for teachers for the high school of this small town on the edge of Sigurd's canoe country. Sigurd was hired, and in February 1923 the Olsons moved to Ely, where Sigurd returned to teaching and searching for a larger purpose.

SETTLING INTO CANOE COUNTRY

Sigurd Olson taught biology and geology classes, first at the high school and later at the Ely Junior College. He engaged his students with organized lectures, dissections, and field trips. Elizabeth and Sigurd had two sons (Sigurd Thorne and Robert Keith), and they shared an active social life. During the summers, Sigurd was a canoe guide for a local outfitting company that serviced the Quetico Park and Superior National Forest area along the Canada-U.S. border. As a guide, Sigurd learned the country and its history, refined his outdoor skills, and developed a more diplomatic personality. Sigurd found solace and a business to run when he purchased part ownership in an outfitting company. He gained exposure to the politics of wilderness protection, made newsworthy at the time by Ernest Oberholtzer and Aldo Leopold. Sigurd wrote articles, published with some difficulty, and began exploring a career as an ecologist, hoping to combine writing with outdoor exploration and research.

At the age of thirty-two, after ten years of teaching, Sigurd uprooted his family to enroll at the University of Illinois at Champaign. His master's thesis was on gray wolf life history and predator control efforts. Sigurd had collected and compiled data on wolves and coyotes the year before, around the woods and lakes of Ely. By examining deer and trapped wolf carcasses and observing wolf habits, he initiated the study of wolves and their prey; his research was published in scientific journals and often cited.

In 1932, the Olsons were back in Ely. Sigurd taught at the junior college and he settled into the routines of family life. A year later Aldo Leopold, professor of game management at the University of Wisconsin–Madison, asked Sigurd if he would be interested in pursuing a doctoral degree in wildlife and ecology. Sigurd was mildly interested and had sent correspondence to accept, but a misunderstanding regarding his employment status eliminated him from consideration. Later Leopold offered Sigurd a wildlife management position, but Sigurd declined the offer. After considerable thought, Sigurd concluded that he did not wish to pursue a career of study and research.

The canoe county wilderness conflicts were brewing, and Sigurd's life changed dramatically after he testified at an International Joint Commission public hearing on a massive proposal to dam and flood extensive areas of the canoe country for economic development. The development was proposed by Edward W. Backus, an aggressive lumber and paper mill tycoon who had built the dams at Koochiching Falls on the Rainy River and Kettle Falls on Namakan Lake. Ernest Oberholtzer had a different vision, one of a world war memorial forest to protect the ten-million-acre Rainy Lake watershed, which included the Quetico-Superior canoe country. Oberholtzer brought Olson into his circle of conservationist friends. They shared the desire for wilderness protection, although Olson had a more restrictive view of what should be allowed within wilderness areas. Olson published articles on wilderness protection and with Oberholtzer's connections found new opportunities for work.

From 1935 to 1947, Sigurd Olson was the dean of Ely Junior College, an active opponent of developments that threatened the wilderness around Ely, a writer of short articles and essays, and a father. Sigurd's older brother Kenneth, the dean of Northwestern University's Madill School of Journalism, encouraged Sigurd to write in the style of a newspaper column. Sigurd found success in this writing style and published many short essays and sketches on nature and wilderness. Sigurd's eldest son, Sigurd Thorne Olson, served in World War II and became a wildlife biologist like his father, studying common loons in Minnesota and fish and wildlife in Alaska.

At the age of forty-eight Sigurd Olson resigned as dean of the junior college and began writing full time. Wilderness protection advocates, however, intervened with an offer to become the Izaak Walton League's wilderness ecologist and advisor to the Wilderness Society. With purpose and tact, Olson dealt with the threat of fly-in fishing in canoe country (at the time, Ely had the most float-plane activity in the United States). Olson interacted with county commissioners, state and federal agency representatives, state elected officials, and U.S. senators and representatives to gather support for a bill that would direct the U.S. Forest Service to purchase private lands in the Superior National Forest Roadless Areas. Olson traveled extensively and wrote timely and moving articles, and six months later President Harry Truman signed the land acquisition bill into law. Next, Olson and his fellow conservationists, including wilderness adventure writer Calvin Rutstrum, worked to have President Truman declare the lower altitudes over canoe country off-limits to air travel for the purpose of preserving wilderness. Again, Olson wrote articles and spoke to local and national groups. In addition, he helped produce a video documentary entitled "Wilderness Canoe Country." The film turned out to be a tipping point for an airplane ban over the canoe wilderness. Shortly thereafter, President Truman signed an executive order that prohibited airplanes landing on lakes and flights below four thousand feet.

In 1956, Sigurd and Elizabeth Olson purchased thirty-six acres on the shore of Burntside Lake near Ely. They named their retreat Listening Point, after the westward facing point on the property. Here, Sigurd contemplated along the shore, watched the sun set across the lake, and touched wildness and nature. The property consists of white pine, birch, juniper, alder, and willow growing on the thin soil. Bedrock extends into the water, and glacially deposited boulders are scattered about the shoreline. Sigurd moved a Finnish square-logged cabin from a nearby farm onto Listening Point and restored it for use as a getaway place. Listening Point was named to the National Register of Historic Places in 2007. Today the cabin on Listening Point and surrounding shorelands are maintained by the Listening Point Foundation, which is dedicated to advancing Sigurd Olson's legacy of wilderness education.

Sigurd Olson's success in protecting the lakes and woods of canoe country catapulted him into the national conservation movement. From 1950 to 1974, he was a leading voice on wilderness and national park issues. He was president of the Wilderness Society and National Parks Association, National Parks Advisory Board member, and active participant in various Quetico-Superior organizations. In his book *Open Horizons,* Sigurd expressed concern about the potential loss of wilderness in canoe country and the march of development elsewhere:

After World War I and during the early twenties, a great road-building program was announced, one which would open up the lake country and make it accessible to tourists. "A Road to Every Lake" was the slogan, and chambers of commerce from nearby communities trumpeted the hope of making the wilderness the greatest resort region of America. No longer isolated, the Superior National Forest would become a mecca for fishermen, "The Playground of the Nation." I read the announcements and editorials with unbelief. Could it be true the wilderness was to be destroyed? Would the lakes and rivers have roads to them all, with cottages and summer resorts lining their shores as they did in Wisconsin, central Minnesota, and Michigan?

Sigurd Olson's work influenced decisions on wilderness across the country, including stopping the Echo Park Dam that threatened Dinosaur National Monument, instituting the National Wilderness Preservation System (1964 Wilderness Act), creating wildlife refuges in

Alaska (including the Arctic National Wildlife Refuge), and developing management plans for many national parks. In his own backyard, he advocated further protections for the Boundary Waters Canoe Area Wilderness (BWCA Wilderness Act of 1978) and authorization and establishment of Voyageurs National Park, and one third of the ten-million acre Rainy Lake watershed that Oberholtzer wished to see as a world war memorial forest was protected. For Olson's leadership on conservation issues across North America, he received the highest awards from the Izaak Walton League, the National Wildlife Federation, the Sierra Club, and the Wilderness Society. At the age of seventy-five, he retired from his professional conservation work.

Remarkably, Sigurd Olson found time to write several books during his twenty-four-year conservation leadership career. His first book, *The Singing Wilderness,* became a bestseller. *The Singing Wilderness* is a collection of wilderness adventures, nature interpretations, and essays. Many of the essays were written while Sigurd was still the dean of the junior college and were rejected by magazine editors at the time. The book is a deep enquiry into the nature of wilderness and into the construction of a sense of place through slow and careful observation, knowledge, and experience. Two years later, *Listening Point* was published. Here Olson wrote about his land ethic and the importance of nature's unknowns, recounting his experiences with his own lakeshore property. Sigurd continued to write, producing five books in the 1960s and early 1970s. He compiled his speeches and talks on wilderness, philosophy, and theology into the book *Reflections from the North Country* (1976). Research he had conducted over a decade earlier on noted contemporary scholars, biologists, and philosophers on the human condition was instrumental in the refinement of his philosophy and religious views expressed in the book. He stated, "what civilization needs today is a cultural sensitivity and tolerance and an abiding love of all creatures, including mankind." Sigurd recognized that deeper appreciation was needed for true respect and effective protection of nature.

On January 13, 1982, a frail Sigurd Olson died while snowshoeing near his Ely home. His last book, *Of Time and Place,* was published posthumously.

LIVING CONNECTED WITH NATURE

Like Aldo Leopold, who tried to recruit Olson to wildlife ecology and management, Olson was aware of the smallness of humans and of their place within ecosystems. He was also aware of humans' power to damage ecosystems and irreversibly destroy a sense of place. He believed, like Leopold, that humans could be educated and that planning and design could play a role in conservation and in restoration of multifunctional landscapes. Olson did not elaborate much on the role of government in planning, but his approach called for citizens and government to tackle conservation issues in unison. Nature is important to humans not only in the form of untouched wilderness, but also in its smaller and more modest manifestations, from park to garden to the creek hidden in agricultural lands. Olson, sensitive to the stresses of modern life, considered nature a tonic. He believed an experience in nature, including some serious physical activity, was a powerful antidote to the overload of stimuli in everyday life. This was not a new idea; Henry David Thoreau and James Muir had shared these sentiments one hundred years earlier. In Olson's time, however, so much nature was perishing that people like him realized something should be done. Simultaneously, science and politics developed an awareness that something *could* be done.

Although, like Leopold, Olson was a product of his time and place, many of his insights into the place of humans in the landscape are still valid. His approach has many implications on the effective creation of a full, rich lakeshore life. Olson was acutely aware of water as the principle of life. The circulation of water through an ecosystem shapes that system more than anything else. Understanding how a landscape works first requires understanding the quality, quantity, and flow of the water. This has been understood in American landscape architecture since Ian McHarg. It follows, then, that a design for a site close to lakes and wetlands should start with an analysis of the hydrology. Olson would not appreciate the practice of simply delineating wetlands or creating buffers around lakes, however useful these practices might be; he would emphasize a deeper understanding of the water system.

Water can also offer a different kind of access to the landscape. Sigurd Olson's canoe not only gave him access to a past of voyageurs and their long-distance routes, it also gave him access to the most beautiful spots in his own region. Much of the variation in landscape and ecology in North America can be experienced best from the water. In areas with dense, privatized shoreline development, the water can be the best way to see the place and experience the landscape. Water routes, therefore, should be an important principle used in designs.

Olson also reinforces the argument that different types of access and different networks of routes must all be considered. A wet area can be perfectly capable of accommodating various land uses, including residential, without jeopardizing vulnerable ecology or identity of place, but only if there is an intelligent combination of slow and fast networks, land and water routes, bike and hiking trails, streets, and roads. Sigurd Olson's Listening Point incorporates his ideas of access in the subtle embedding of small pedestrian paths in the natural landscape and the interweaving of pedestrian and canoe routes at the greenstone ledge sitting at the water's edge. Olson felt guilty about the construction of a driveway to Listening Point but accepted it as a necessity.

Olson acknowledged the need for human use of places, as he acknowledged the need for smart combinations of land uses. He was disturbed by the whistle of the train shattering his wilderness experience but quickly realized that human appreciation of wilderness would not be possible without access, without technology, and without a level of comfort made possible by a history of economic development. Like his cabin, homes and cabins can be embedded in the landscape, streets and trails can be carefully designed, and a rich and varied experience of the landscape can be created and preserved.

Olson was supremely sensitive to the variety and continuous flow of events in a landscape. Looking at the land like he did can help planners and landscape architects become more aware of a place, of what is specific to that place, and of its identity. Olson's interest in identity at different scales is also critical. From the Quetico-Superior, to the site of Listening Point, to the microlandscapes of ancient lichen-covered boulders that fascinated him so much, Olson was always attentive to the unique character of a place. An obsession with regional identity can overlook local qualities, and total immersion on the site level can make us forget qualities or issues at a larger scale that should not be ignored.

The unifying spatial identity in both Olson's work and in this book is *The North*. The North is a mythological construction that is also very real. The lure of northern lakes is partially the result of Olson and other writers capturing the essence and identity of these places. It is partly thanks to writers like Olson that people are aware of the beauty and unique character of wild lakes. Olson and other great writers can help others reframe their picture of the landscape and change the image of desirable landscapes. Appreciating and understanding these areas like Olson did can guide our decisions on the future of many North American lakes.

Sense of place cannot be disconnected from time. Time gives character. A place without a history may be harder to grasp. Of course, there is always a history. Olson saw history everywhere, in multiple layers and manifestations. Human history, like that of the voyageurs, often leaves visible marks like the rock paintings Olson fondly described. It also leaves invisible marks, as stories of the past add interest and individuality to a place. History is also ecological history. Both humans and animals leave traces. Geological history often shapes a place on a grand scale. Olson's beloved Quetico-Superior is largely the product of ice-age glacier movement, revealing and molding a much older volcanic substrate. Ely greenstone, 2.7 billion years old and once believed to be the oldest exposed rocks on earth, is a source of local pride and local identity. Time can also be the time spent in a place. Time produces familiarity, attachment, or personal histories that tie people to a place. Over time, people change places, but places can also change people. A heightened awareness of a place, or a long familiarity with it, can shape the character of a person in ways that are not always perceptible to himself. Designing becomes a more delicate undertaking, seen in that light. Shaping environments is shaping the people living in those environments.

Although the scales of time can be vast, the seasons are the cycle most observed on the scale of everyday life. For Olson, a sharpened awareness of the environment came easily with a heightened awareness of the seasons. *The Singing Wilderness* follows a cycle of seasonal changes, and Olson described the seasonal changes of plants, animals, water, and the visual qualities of a place. He shared access to the full range of features, events, and relations that mark any given place. With a very natural eloquence, Olson advocated for a full experience of nature and landscape through all the senses.

His description of a canoe excursion to the edge of a cranberry bog in fall, the physical difficulties of the terrain, the pungent smell of the peat moss, and later, the experience of cooking and the tart taste of the cranberries, is unsurpassed in its capturing of character and in its exhortation to go outside and experience this yourself. After reading this sketch, who could be insensitive to damage to a site like this? For landscape design, even a vignette about a cranberry bog can have wide ramifications. Preserving places and activities like those Olson described, giving access to these areas without destroying them, and retaining the character of a site while understanding its utility for other activities become major challenges. Unfortunately, many routine practices of development would overlook the bog, overlook the potential for framing it in a design and in subtle routings, and overlook the potential for rich experiences in small sites.

Design as Olson would likely encourage, therefore, is designing for a sense of place. This place can only be grasped through slow and careful observation and, more broadly, experience. Site visits and site analysis, therefore, should be considered much more than a procedural requirement. Walking or paddling around the site is a good place to start. A sensitive site analysis will dramatically improve the quality of the design plan. Designing for identity, unsurprisingly, takes more time than standard development procedures. In the long run, however, a neighborhood respectful of nature and landscape, a neighborhood with a distinct character, will prove more stable in value and stronger in attraction.

Sigurd Olson's work rarely references designs for areas with higher densities but does overflow with references to the need for nature and identity in everyday life. Wilderness cannot be restricted to faraway places. Infusing a spirit of wilderness into heavily used areas around water is an immense challenge. Olson's thoughts and writings provide inspiration on how to meet this challenge. He envisioned an asset-based approach to design for lakeshore living. Here, creating unique places depends on building on existing qualities of the site. Interventions in the landscape are executed with surgical precision; projects should be embedded in the landscape and build on existing features, objects, and structures. Working with the site is working with nature, not against it. In many cases, this makes projects not only more attractive and sustainable, but also less expensive. Asset-based designing first involves a careful inventory of assets. This inventory cannot be conducted indifferently, but can only follow from knowledge, time, and a sensitive approach to the site. It also involves careful routing: networks of motion must capitalize on fragmentary spaces that already lend themselves to walking, paddling, or driving; connect interesting objects and views; and skirt vulnerable places and creatures. Fragments can be connected to form ecological networks that allow people and animals to move. Roads and streets can be inserted in a landscape. If placed carefully, roads can separate people and animals where necessary, but streets can combine and overlap them where possible. Along the routes, areas can be created that provide space for various activities, and assets can be framed to make the natural experience more intense.

Asset-based designing is not a purely technical activity, because assets are in the eye of the beholder. Views can be important assets of a site, but if they are not noticed, recognized, or valued, they cease to be assets. More subtle experiences may be even harder to appreciate. Raising awareness is again the key. Olson's works can help heighten the awareness of planners, owners, or politicians, inspiring them to observe and appreciate a site more acutely. They will all see more, and differently, after reading Olson.

In several places, Olson reflects on portages and jumping-off places. He talks about those special places in both a literal and a philosophical sense. For the canoeist, the portage is a place of physical labor, but also a place to enjoy the surrounding landscape. Portages can be old or new, clearly visible or quickly overgrown. They act as connectors, joining two lakes or streams on a long voyage. Jumping-off points are where the journey begins; they are places to gather, to look ahead, to reflect and take a step into the unknown. Both portages and jumping-off points are places of transformation. They offer a change in experience, and perhaps even a change in self. Portages and jumping-off points can be incorporated in everyday life. They can take the form of special spots for reflection or areas with staged shifts in experience. These places often become special in a way that other places do not. One of Sigurd Olson's prevailing beliefs was that wilderness can be close to home. He wrote most of his works in a converted detached garage a few feet away from his home. The garage, his "writing shack," was his jumping-off point. A minimal physical distance gave him the distance he needed for reflection and a separation from everyday hassles. It opened up new horizons for him. In this shack he felt part of nature; on a desk sat a collection of rocks, and behind the garage stood a wall covered with lichens and mosses. This one place allowed him to work, to think, and to play. Good designs should accomplish the same. It should create places that can accommodate, invite, or trigger a variety of activities, experiences, and interpretations. Planning and design for unique place identity cannot be authoritarian; it should not try to instill a place with one meaning, one experience, or one function. Plans can guide movement, but they should always leave room for exploration. They should encourage people to take a moment, to look at things differently, to explore new perspectives. That way, not only can people better understand the character and uniqueness of the place, but the place can change people for the better.

CHAPTER 5

William Whyte and Human Habitat

William Hollingsworth "Holly" Whyte Jr. was a journalist and sociologist. His work spanned numerous disciplines, including land use, rural and urban development, sociology, regulations, and public policy. He is perhaps best known for his bestselling book *The Organization Man,* in which he documented the diminishing individualism in American society. He was also an astute observer of people, and his description of people's behavior in urban settings revolutionized metropolitan design. Whyte was a humble and optimistic man, not prone to overstatement, and he had a preference for action. Holly Whyte had a unique gift in communicating his detailed observations and his insights on what made a place more livable or civil. He foresaw the decline of cities—the consequence of cheap energy, federal programs, and the social marketing of a new vision of America centered in suburbia. But by the end of his life, he witnessed a reversal of that trend and the fortune of cities rebounding, in part by his work and the others that followed to pursue development of more humane cities. The places he came to love and his proposals and strategies to protect and enhance our habitats, both in the country and within the city, are scattered throughout his writings. Through his insights into human behavior and his straightforward and commonsense principles of urban space design, William Whyte's ideas continue to have a profound influence on land-use practices and urban design today.

TOWN, COUNTRY, AND SHORELINE

William Whyte was born on October 1, 1917, in West Chester, Pennsylvania, twenty-five miles west of Philadelphia. William, son of a railroad executive, spent his youth both in town and in the rich, rolling hill countryside of the Brandywine Valley. The landscape of his formative years made a lasting impression, as he would work some forty years later on promoting tools to manage urban sprawl within the Brandywine Valley and across the continent.

The town of West Chester is located on the upland clay and loam soils between two drainages. West of town is Brandywine Creek and to the east lies Chester Creek, both flowing into the Delaware River. The area is rich with early American history. The country was settled by Swedes and English Quaker immigrants, who displaced the Algonquin Indians (Lenape). In 1777, before the town was established, George Washington fought and lost the largest battle of the Revolutionary War. The loss was attributed to poor understanding of where the enemy was and unfamiliarity with the territory. About 150 years later, a young William Whyte took excursions to the same countryside, enlightened by a sense of history. The town of West Chester was established at the turn of the nineteenth century. Interestingly, the county courthouse in West Chester was designed in the 1840s by Thomas U. Walter, who would later design the U.S. Senate and House wings as well as the central dome of the U.S. Capitol. The Pennsylvania Railroad, nurseries, and industry employed many of the residents at the turn of the twentieth century, and few residents worked outside of town.

William Whyte grew up in the aftermath of World War I, and he witnessed the Great Depression as a teenager. Yet, at the end of his life, he looked back fondly on those times. He recalled playing war games with his cousins, using air rifles as weapons and trenches for cover. He spent his days exploring West Chester and the nearby countryside. During his youth, the town was compact and getting out to the country was easy. Whyte noted, "The country began exactly where the town ended." He spent much of his time at his Grandmother Whyte's home, and he and his friends often biked the trails and roads of the Brandywine Valley. William also spent time at the Chester Valley farms, owned by his Grandfather Price, and at Grandmother Price's home on Cape Cod. The home was on the shore of a pond in Wellfleet, Massachusetts. William picked blueberries and blackberries alongshore, and he knew the paths and the beaches. He made maps of the area that he shared with his friends and family.

William was a bright student. At age fourteen, he was sent to St. Andrew's School, a secondary education boarding school on 2,200 acres in Middletown, Delaware. On the newly founded campus, William enjoyed the pleasures of life on the shores of Noxontown and Silver Lakes (a reservoir of the Appoquinimink River). He attended grades eight through twelve and later recalled that he learned how to write from Bill Cameron, a gifted English teacher at St. Andrew's. William served as editor for the *Cardinal,* the school newspaper, where Bill Cameron instilled a crisp and direct writing style devoid of similes, slang, and metaphors.

The land, water, and shores of William Whyte's youth, like most areas, continue to face development pressures. Presently, the Brandywine Valley has retained much of its character due to the efforts of citizens working for the Brandywine Conservancy—more than forty-three thousand acres have been protected and preserved with conservation easements (i.e., land preservation agreements). Cape Cod National Seashore was created in 1961 to preserve a portion of the peninsula, the longest expanse of sandy shoreline on the East Coast. Today, the Appoquinimink River, which flows through the St. Andrew's School campus, is polluted and listed as impaired water, as it receives considerable nutrients from stormwater runoff. The Appoquinimink River Association, St. Andrew's School, and lakeshore residents have begun to revegetate the shoreline to reestablish shoreline buffers, and the Town of Middletown adopted a riparian buffer ordinance. In addition, education efforts are underway to change the behavior of people in the watershed to minimize rainwater runoff and protect natural areas.

Upon graduating St. Andrew's School, William attended Princeton University. He was on the editorial staff of the *Nassau Lit,* the college magazine, and wrote a prize-winning play. He graduated cum laude in June 1939, with a major in English. After graduation, William elected to attend a management-training program at the Vick Chemical Company. Euphemistically named the Vick School of Applied Merchandising, it was actually a disguised salesman-training program rather than an executive management program. William worked as a salesman for Vick's products to Kentucky country storekeepers. Years later, he wrote an amusing story for *Fortune,* titled "Give the Devils No Mercy," about the sales strategies and corporate culture at the time of his employment. World events and his own drive and sense of responsibility resulted in him leaving his career as a salesman.

In 1941, at age twenty-four and before the United States declared war on Japan, William Whyte joined the Marine Corps. He attended Officer Candidate School in Quantico, Virginia, as a candidate for second lieutenant. William believed that the United States soon would and should be at war, and he wanted to serve "with the one truly elite outfit in the U.S. armed services." On August 7, 1942, eight months after the attack on Pearl Harbor, William Whyte was standing on the beach of Guadalcanal, a large tropical island in the South Pacific, as a first lieutenant and battalion intelligence officer. Guadalcanal was the first serious engagement between U.S. and Japanese ground forces (and the first American ground offensive of the war). For four months Lieutenant William Whyte of the Third Battalion, First Marine Regiment, First Marine Division, would be educated in the misery of war. As battalion intelligence officer, Lt. Whyte conducted patrols and scouting missions. He provided enemy soldiers for interrogation, and he conducted analyses of captured Japanese documents, including soldier diaries. There he witnessed numerous acts of bravery and stupidity. Lt. Whyte left Guadalcanal with malaria in December 1942. He spent time recuperating in a hospital in Tasmania and camps in Australia. For a short time he taught a map-reading course to Marines at a scout and sniper school, but reoccurring bouts of malaria forced Whyte to return home. Whyte returned to West Chester and spent several months recuperating from his illness. In September 1943, Whyte was back at Quantico to teach Marines about the Japanese military and soldiers. Whyte's talent for observation, powers of perception, and ability to draw meaningful conclusions proved valuable.

Prior to the end of the war, Captain Whyte wrote a series of seven articles in the Marine Corps *Gazette,* the professional journal of the U.S. Marines, including an article on the Guadalcanal Campaign from the Japanese perspective. In the articles, Whyte stressed the importance of teamwork for effective intelligence gathering, and he emphasized the need for analyses of enemy capabilities and continuity in intelligence-gathering procedures. Whyte reminded Marines of the following: "The transformation of information into intelligence requires no super-mind gifted with a mystical intuition, but one with sufficient humility not to short-cut sound procedure and jump to immediate conclusions." This lesson is one that William Hollingsworth "Holly" Whyte also lived. He would later use an analytical approach to study suburban society, land use, and urban open spaces in exquisite detail, much like an anthropologist studying the Inca site of Machu Picchu. His mission was to apply his research to beautify, preserve, and humanize city and countryside. Whyte said, "It takes great patience to observe things accurately . . . a common human frailty: we see what we want to see and we hear what we want to hear." The English major transformed himself into a scientist out of necessity of war and a personal desire to succeed in understanding the world around him. Whyte asked and attempted to answer questions that few planners and developers had asked before.

SUBURBIA

After the war, Holly Whyte retired from the Marines. He worked for thirteen years as a writer for *Fortune* magazine, rising to assistant managing editor. Whyte was a productive writer and editor, and his observational skills and ability to capture both micro- and macroperspectives into words were valuable. His first major story, "The Class of '49," published in 1949, was about the postwar cohort of college graduates. The postwar years were a dynamic period; the

Great Depression was over, and returning soldiers were taking advantage of the Servicemen's Readjustment Act of 1944 (i.e., the G.I. Bill) that provided college and vocational education. A massive change in land-use patterns and lifestyle, underway since the 1920s, was accelerating in the 1950s: suburbanization. Holly Whyte had a passion for the beautiful countryside of his youth and a growing awareness of the importance of the city, and this wide-ranging phenomenon caught his full attention and interest. Gasoline had just become inexpensive, and the supply was abundant; soon the Federal Aid Highway Act would appropriate money for the construction of the U.S. interstate system. The U.S. Federal Housing Administration loan program increased suburban home construction. A car-dependent lifestyle was emerging for both the country and the city. Streetcars, the efficient movers of people within cities, were beginning to disappear; the last streetcar functioned in Minneapolis in 1954, and Wisconsin's electric streetcars ended operation in 1958. In this context, William Whyte began investigating the suburbs.

In 1953, *Fortune* published an article by Whyte on his observational studies of social interactions of residents of Park Forest, Illinois, a newly created suburb and community. The 2,400-acre Park Forest community was built in cornfields, just thirty miles from Chicago. The development consisted of clusters of rental apartments, a central shopping center, and single-family homes on 60-by-125-foot lots arranged on curved superblocks at the periphery. Whyte and his *Fortune* associates also investigated the suburban communities of Drexelbrook and Levittown, Pennsylvania. These were intensive sociology studies. Like a military intelligence officer scouting enemy movements, Whyte mapped activities from social events, religious engagements, and civic leadership to the webs of friendship of both adults and children. He noted, "we mapped it all, and my maps revealed fascinating patterns that couldn't be attributed to chance." The distribution of activities was not random; the patterns were consistent with a clumped distribution, and the patterns persisted with resident turnover. Whyte was intrigued by these studies and believed that, given the pace of suburban development, suburbia would have a profound influence on the social life and values of America. In addition, Whyte's analyses demonstrated the importance of the physical layout of a community in determining neighbor interaction and civil engagement. He identified several key physical factors that had consequences for linkages between families. Whyte stated, "Despite the fact that a person can pick and choose from a vast number of people to make friends with, such things as the placement of a stoop or the direction of a street often have more to do with determining who is friends with whom." The key physical factors that Whyte identified included children's play areas, placement of driveways and porches, adjoining yards, centrality of home, chronology of construction, and physical barriers and boundaries. These physical factors, sometimes quite subtle, were mainly the result of development decisions.

Whyte's writings for *Fortune* on the culture of postwar America eventually evolved and expanded into *The Organization Man,* published in 1956. *The Organization Man* presents Whyte's analyses of suburbia and his thoughts on the abdication of individual responsibility to organizations. He asked readers to think about how their own environment impacted their lives and whether the new social pressures to belong to a group diminished their independence, creativity, and responsibility. The book became a bestseller with millions of copies sold and was translated into many languages. After publication of *The Organization Man,* Holly Whyte, now a senior editor at *Fortune,* began planning investigations and magazine coverage on urban sprawl and American cities. In 1957, Whyte facilitated a discussion on the plight of cities with nineteen planning and urban experts. He planned a series of six essays that highlighted the problems faced by cities. American cities were in decline, a result of the suburbanization phenomenon and downtown redevelopment projects that forced suburban shopping center styles and carcentric dependences into downtowns poorly and inappropriately. While the essay series primarily dealt with cities, Whyte recognized that urban issues could not be separated from suburbanization. He stated, "uncontrolled development on the fringe of the metropolitan area is going to extend its limits vastly—and ruin much of it at the same time." Whyte contributed two essays to the series, one on American cities and one on urban sprawl.

In his urban sprawl article, Whyte presented the first problem not as land consumption or conversion of farms or natural areas to residential development, but rather as lack of an appropriate pattern of growth. Growth patterns were almost entirely determined by builders and developers, without giving thought to aesthetics, environment, health, macroeconomics, or even functionality. In addition, he noted that shorelines were quickly being developed and open beaches were disappearing. Whyte provided readers with examples of communities that effectively addressed sprawl with tools that included open-space plans, regional planning, land acquisition, and zoning ordinances. These communities took back responsibility from the builders and developers in directing the pattern of growth and the form of the community. This essay drew the attention of Americans back to the forgotten need for planning. As Whyte himself noted, he outlined a conservative approach. He stated: "Where the new developments are scattered at random in the outlying areas, the costs of providing services become excruciating. There is not only the cost of running sewer and water mains and storm drains out to the Happy Acres, but much more road, per family served, has to be paved and maintained. Who foots the bill for the extra cost of services? Conventional tax practice spreads the load so that those who require the least services have to make up the difference." Whyte discussed land acquisition options for open space including outright purchase, purchase of development rights, and land donation programs. He cited examples like the Massachusetts Trustees of Reservations, who today have now preserved over one hundred properties and historical sites on twenty-five thousand acres of land. Whyte outlined the second problem of suburban development as "how to achieve an economically high density in developed areas and at the same time more amenable surroundings for the people in them." He suggested controlling the locations of subdivision developments and of the layouts within the subdivision. Although Whyte didn't invent the concepts of purchase of development rights and cluster development to protect the countryside, he spent time refining these concepts, and his essays, other writings, and promotion of these two tools would greatly expand their use.

The six essays were assembled into a book, *The Exploding Metropolis,* which was published in 1958. Other contributors included Francis Bello on mass transit and Jane Jacobs with a sharp critique of downtown redevelopment. (Jane Jacobs received a grant from the Rockefeller Foundation to expand her critique of modernist planning and design on downtown communities into a book, and in 1961 *The Death and Life of Great American Cities* was published.) Whyte was a supporter of mass transit, predicting in the book's introduction that cars would result in "no nice clean regional towns, but a vast sprawl of subdivisions, neither country nor city." Whyte also concurred with Jane Jacobs. He labeled many downtown redevelopment projects as "anti-city," describing them as "sealed off from the surrounding neighborhoods as if they were set in cornfields miles away." These developments, "designed by people who don't like cities," resulted in the banishment of people from street and sidewalk. In 1957, Whyte called for the design of city habitats based on the pedestrian rather than the driver.

The Exploding Metropolis set a new course for Holly Whyte. He wanted to expand his work on urban sprawl by alerting people about the problem and assisting governments with finding solutions. In 1958, Whyte took a leave of absence from *Fortune* to work on a technical bulletin for the Urban Land Institute on the concept of purchase of development rights. He was supported by the generosity of Laurance S. Rockefeller and the money he received from the sale of *The Organization Man*. In 1959 Whyte completed the technical bulletin. In the report, entitled *Securing Open Space for Urban America: Conservation Easements,* and in an article for *Life* magazine ("A Plan to Save Vanishing U.S. Countryside," August 1959), Whyte advocated the use of conservation easements as a development tool. Although Whyte did not invent the concept of conservation easements, they were poorly understood and rarely used at the time. Whyte's work was credited with open-space legislation enacted in California, Connecticut, Massachusetts, Maryland, New York, Pennsylvania, and elsewhere. By the 1980s conservation easements were used widely. For example, groups like the Mount Vernon Ladies' Association protected the shoreline of the Potomac, and the Minnesota Land Trust has permanently protected over one hundred and sixty miles of Minnesota shoreline. Upon completion of his work promoting conservation easements, Holly Whyte continued to work on the loss of America's countryside caused by urban sprawl. He became an advocate for open-space protection. He lobbied to members of the American Society of Planning Officials, the Urban Land Institute, and the American Society of Landscape Architects. In 1961, Whyte gave testimony at a hearing at the U.S. Senate on a bill for open-space protection.

In the 1960s Whyte worked as a consultant for the U.S. Outdoor Recreation Resources Review Commission. The commission was comprised of eight members of Congress and seven citizens appointed by President Dwight D. Eisenhower, and its task was to recommend actions for conservation and outdoor recreation to the president and Congress. Chaired by Laurance Rockefeller, the commission began with a comprehensive review of facts on outdoor recreation. It contracted with universities and research organizations for detailed studies on water recreation, wilderness, and hunting and fishing. The commission surveyed outdoor resource managers and citizens, projected future demands, and detailed potential actions critical to ensure that Americans would have outdoor resources into the future. One primary finding of the surveys was that lakes and rivers were the focal point for much of the country's outdoor recreation. In January 1963, the commission submitted its ~4,800-page report to President John F. Kennedy and Congress. William Whyte contributed to the centerpiece of the report, and he authored volume 15, *Open Space Action.* Whyte recognized the aesthetic, social, and ecological service value of open space, and in his volume he called for action to acquire and preserve open space for the benefit of citizens and for the greater use of cluster developments over "the land-wasting pattern" of the conventional subdivision. The commission's report was comprehensive and broad in scope, and it outlined a five-point set of recommendations to address outdoor recreation into the future. The recommendations consisted of the development of a national recreation policy, a classification for outdoor recreation, expansion and intensification of outdoor recreation programs, an agency to lead planning efforts, and a program of grants-in-aid to the states. The report was the foundation for future landmark environmental legislation such as the Wilderness Act, the Land and Water Conservation Fund, the National Trail System Act, and the Wild and Scenic Rivers Act. Tens of thousands of projects have been completed and millions of acres of land have been acquired or protected since these laws were created.

Holly Whyte would continue to work with Laurance Rockefeller. The Rockefeller Brothers Fund supported Whyte's work on urban land issues and provided him with an office in the Rockefeller Center, New York. Whyte's book *Cluster Development* was published in 1964 by the American Conservation Association with a foreword by Laurance Rockefeller. In the book, Whyte expanded and elaborated on his early thoughts about conventional subdivision development. Whyte later noted, in *A Time of War:*

> In the suburbs, we had these little lots filled up with Cape Cods, ranches, and Hansels and Gretels, gobbling up land at a prodigious rate . . . we came up with an ancient concept—grouping houses closely together and using the land we saved to create common greens and squares. We called it cluster development. And it worked. . . . What we found was that people like to live in these open space developments, once they get used to watching their children play near streams, valleys, and woods. Once again, it was a simple idea, growing out of the way we observed land patterns.

In 1964 Holly Whyte married Jenny Bell Bechtel, a fashion model and clothing designer. Together they had one daughter, Alexandra. Despite his new role as a family man, Whyte remained dedicated to work on open space and urban development and found time to be active in the Municipal Art Society of New York.

In association with Laurance Rockefeller, Holly Whyte was the manager of President Lyndon B. Johnson's White House Conference on Natural Beauty (1965), which was primarily funded by the American Conservation Association. This private nonprofit organization was founded to advance understanding of conservation and to preserve beautiful natural places. (Interestingly, A. Starker Leopold, son of Aldo Leopold, was a 1965 conference steering committee member for the organization.) Whyte also served as a member of President Johnson's Citizens' Advisory Committee on Recreation and Natural Beauty (1966). Whyte drafted the advisory committee's final report and included his idea of a tree-planting program for urban beautification. In addition, Whyte chaired New York governor Nelson Rockefeller's Conference on Natural Beauty. In the late 1960s, Whyte took his work for the American Conservation Association, the Outdoor Recreation Resources Review Commission, and the Rockefeller Brothers Fund and began to write a book to bring together his ideas on the importance of open space and the tools to help protect it. The book became highly influential and today is a classic study on land use.

This book, *The Last Landscape,* was published in 1968. It is a thoughtful and easy-to-read book that captured Whyte's ten-year period of intense study of land use and his uniquely voiced call for immediate action. Whyte wrote philosophically but also practically, giving advice on open-space protection and responsible shoreland development and suggesting strategies to oppose poor-quality developments. His message of urgency and optimism is refreshing and contagious, and his solutions remain vital today. Early in the book, Holly Whyte discussed the tools of zoning, land purchase, and conservation easements. He examined the strengths and limits of these tools, and he exclaimed, "parochialism is the great weakness of local zoning." For example, it is still common for many local governments to rely on large-lot zoning provisions to protect open space and the character of their community. Whyte noted that large-lot zoning does not save natural resources or open space. Large-lot zoning eats up more land, and the gains are often short-lived as economic and social pressures minimize the benefits of this technique to protect shorelands. The pressures to split large lots are very strong, and people who own large lots often request lot splits as property values rise. Whyte stated:

Large-lot zoning does not save open spaces. It squanders it. By forcing developers to use large lots for little houses, the community forces them to chew up much more of the open landscape than they have to. Instead of several tightly knit subdivisions, housing will be spattered all over the place.

The true purpose of large-lot zoning, of course, is to keep people away, or at least people with incomes lower than the [local] majority. It can work very well in this respect, but over the long run the victory is likely to be pyrrhic. The community may shunt developers and middle-income people away, but they will soon fill up the interstices, and the surrounding landscape that the community took for granted will disappear. The community won't be penetrated; it will be enveloped. . . .

The fact remains, however, that the ultimate force for knocking down large-lot zoning is not the developer; it is the pressure of people looking for a place to live.

In the long run, this will [win] out. It is quite possible that in time we may have ordinances for maximum lot sizes. If the community is reconciled to housing more people and wants to save open space, it will have to reverse course and demand that the developer use less land for putting up his houses. In the form of cluster development, steps in this direction are already being taken.

Whyte believed that only a small number of forms of zoning worked effectively. Those include floodplain zoning, cluster zoning (e.g., planned-unit development, cluster development, or conservation subdivisions), and historical district zoning. Other forms of zoning, such as agricultural or open-space zoning, are notorious for their failures.

In *The Last Landscape* Whyte again focused on cluster developments, revisiting and developing the ideas of his 1964 book. Whyte asserted that cluster developments were actually a "counter-revolutionary movement," as this land-use pattern was old, tested, and proven. Whyte wrote, "it is the principle of the early New England town; it is the principle of the medieval village; it is, in fact, the basic principle of community design since we first started building several millennia ago." Whyte was critical of the new subdivision form—the conventional lot-and-block design, where lots were placed without regard to the land features and with a cynical contempt for the local ecology. Whyte noted, "Subdivisions are named for that which they are about to destroy." He gave advice on design of cluster developments and acknowledged some of the problems, such as the need to demarcate with fences or hedges the line between the private open space and the common open space; private space is important, and a clear designation encourages appropriate use of both private and common spaces. He noted that these developments do not require public entities or charities to establish open-space areas for neighborhoods. He stated: "For years, planners had been arguing that if lots took up less of the land, subdivisions would be more economical to build and more pleasant to live in. Rather than divide all of a tract up into lots, they suggested, developers could group the houses in clusters, and leave the bulk of the land as open space. It was an ancient idea: It was the principle of the New England village and green, and its appeal had proved timeless." He also argued: "The only possible way we can save much open space is to use every tool we can get our hands on and use them together. There has to be a unifying plan, and we must be as hard-boiled as the speculator in framing it. We must identify what cannot be saved, what can be and should be saved, and tackle the job as though there will be no reprieve."

Whyte also addressed the creation of new towns, some of which he thought were overly planned anticities, as these developments often neglected an important urban principle—"The essence of a downtown is concentration and mixture. . . . Urbanity is not something that can be lacquered on; it is the quality produced by the great concentration of diverse functions and a huge market to support the diversity." Whyte understood that this required attention to details at the human scale. Whyte also argued for the efficiencies of concentration and the importance of the city in protecting the landscape. He called for open-space standards, with aesthetics and linkage between natural areas as critical considerations.

Whyte closed the book with a call to action: "So let's be on with it. . . . If ever there was a time to press for precipitate, hasty, premature action, this is it. . . . We have no luxury of choice. We must make our commitments now and look to this landscape as the last one. For us, it will be."

After the publication of *The Last Landscape,* Whyte continued his work on open-space protection and also got involved with broader environmental issues. In 1970 and 1971, Whyte contributed to reports for President Richard M. Nixon's Citizens' Advisory Committee on Environmental Quality. These reports dealt with pollution reduction, energy use, environmental education, and how to fund environmental programs. Whyte also wrote a guide on citizen participation in environmental issues entitled "Community Action for Environmental Quality." The guide was widely distributed across the country, and based on this work the American Broadcasting Corporation developed a three-part television series on the issues and solutions outlined in the guide.

CITIES

During the 1960s Holly Whyte was also involved with planning and zoning in New York City. Whyte was hired by the New York City Planning Commission to revise the city's comprehensive plan. The master plan was completed in 1969, and the *New York Times* and the American Society of Planning Officials praised it for its clarity and vision. This work exposed Holly to the city's zoning ordinance, including the 1961 incentive zoning provision that gave developers added floor space with creation of small privately owned public places. The value of these required places intrigued Whyte, as the city administrators had not evaluated the merits of this incentive ordinance. While the plazas and other spaces were clearly worth it to the developers (they were receiving a large multiplier on their public place), was the public benefiting from these spaces? Whyte worked to answer that question.

In 1970, Whyte accepted a one-year term as Distinguished Professor of Urban Sociology at Hunter College of the City University of New York. He engaged students to assist him in his studies on city public places, and he received numerous grants to fund this research, including an "expedition grant" from the National Geographic Society to conduct an anthropological study of New York City's inhabitants. Whyte wrote in *The Exploding Metropolis:* "I found in my work on urban spaces that many of the most rudimentary questions were neither posed nor answered. The effect of sun on siting, for example; wind and other micro-climatic factors. The customary research plans didn't help much because the research was vicarious, that is, once or twice removed from the reality being studied. There is no substitute for a confrontation with the physical. You see things that theory misses."

Whyte's study, the Street Life Project, provided a thorough analysis of the behavior of New Yorkers in public places. Whyte and his colleagues conducted interviews and surveys, produced maps and diagrams, and used time-lapse photography. Whyte observed that small intimate places were busy places compared to broad, large-scale plazas and places near large blank walls.

He also found that public spaces need copious amounts of sitting places (e.g., ledges and movable chairs) along with tables, water, trees, sun, and food and retail nearby. He studied the details of each of these amenities and how they were best combined to produce quality places that attracted people. Whyte's research was instrumental in the revision of New York City's incentive zoning ordinance in 1975. In addition, based on this work, many North American cities adopted similar provisions in their zoning ordinances. This long-term pioneering research resulted in Whyte's 1980 book and film titled *The Social Life of Small Urban Spaces,* and his 1988 treatise book *City: Rediscovering the Center.*

Whyte came to deeply believe that compact, mixed-use development made human places most livable and vibrant. In his late sixties and early seventies, Holly Whyte continued to provide practical advice to numerous large city planning and zoning officials, including those in San Francisco, Dallas, Kansas City, Seattle, and San Diego. In 1984 New York mayor Ed Koch awarded Whyte with the Doris C. Freedman Award in recognition of his long-term research on public places, and in 1985 Whyte was awarded an honorary membership in the American Institute of Certified Planners. Today the organization Partners for Livable Communities presents outstanding individuals working to achieve livability with the William H. Whyte Award.

PRINCIPLES FOR HUMANE LAKESHORES

Holly Whyte believed in certain general principles related to development. He believed in something akin to the broken windows theory for landscape beauty. He observed that small elements can have "a large leverage effect," such that the sum of small trivial projects can have a major impact on the perception of the environment. Whyte also was an advocate for an adaptive management approach. He was not fond of the saying attributed to urban planner Daniel Burnham: "Make no little plans. They have no magic to stir men's blood and probably will not themselves be realized." Despite its grandiose implications, Whyte thought it reckless to undertake large developments without taking time to adapt to both successes and failures. Whyte also understood that developments are planned and designed for humans, and therefore details must be presented at *the scale of human observation and experience.* Perspectives from other points of view provide little help.

Whyte outlined four general principles in *The Exploding Metropolis* necessary for success in creating humane environments, including lakeshore environments:

1. Getting something done is primarily a matter of leadership, rather than research.
2. Bold vision, tied to some concrete benefit, can get popular support fairly quickly.
3. The most effective policy is to get the land first and rationalize the acquisition later.
4. Action itself is the best of all research tools to find what works and what doesn't.

From *The Last Landscape,* one can add:

People are stirred by what they can see. And they are right.

We are going to have to work with a much tighter pattern of spaces and development, and that our environment may be the better for it.

Transportation follows transportation.

With regard to open space: "linkage is key."

Finally, from *City: Rediscovering the Center,* one can add:

Open space, like development, needs the discipline of function. Use it or lose it.

The social life of the street is critical: "It is the river of life of the city, the place where we come together, the pathway to the center. It is the primary place."

Whyte had a passion for the countryside and the city. He worked to create better places to live. More efficient and community-oriented land use outside of cities is, at the moment, constrained by reliance on the automobile, but Whyte believed that a civil society with a high quality of life required anthropologist-designed spaces and planned communities. His landscape planning and design principles, as well as his gift of writing and ability to effectively work with government officials, were instrumental in his success in helping create places that people continue to love and protect today. Whyte's views on city streets can still be heard in the messages from today's progressive city planners on how and why to design streets for people first rather than first for vehicles. In addressing today's problems of place, there is much value in revisiting Holly Whyte's work and taking in the insights of the quintessential Observation Man.

PART THREE

Lakeshore Development and Redevelopment

WE MUST TREAT OUR LAKES WITH GREAT CARE. IF WE SEE A BETTER FUTURE AS A SUStainable future, and if we believe that humans and the rest of nature can live together around lakes, we need capable planning, design, and policy. Leopold, Olson, and Whyte illustrated the need for careful site analysis in order to grasp the identity of places and their embeddings in nature and culture. They also understood the potential fullness of the experience of lakeshore living and the potential richness of place identity. We need the tools to preserve and create environments that allow for such full, rich lakeshore living and people who recognize and value that potential. Several principles for sustainable and just planning and design, advocated by Leopold, Olson, and Whyte and supported by an understanding of lake ecology, are immediately clear.

First, design for rich lakeshore living must be asset based. A strong awareness of the existing features, qualities, and threats will be necessary to maintain or enhance place identity and accommodate people and nature. Asset mapping, such as site and issue analyses, and asset preservation will have to be part of the recipe.

Second, asset mapping must inspire asset creation. This design process will allow the creation of new spatial qualities. In many cases, shoreland development will have to be redevelopment, and the definition of lakeshore living will have to be expanded.

Third, connectivity must be an essential concern. Ecological networks, water systems, transportation connections, and community design revolve around connectivity or the lack thereof. Connectivity becomes even more important with the realization that the final design will have to accommodate a mix of land uses, in ways not often envisioned in standard development practice.

Sustainable design, inspired by Leopold, Olson, and Whyte, can enable the coexistence of different creatures, different uses, and different experiences. Through the mapping and design of networks and connections, and the understanding of the ecological, hydrological, cultural, and visual qualities of a place, we have the opportunity to truly experience rich lakeshore living.

Asset Preservation

THIS BOOK IS ABOUT LAKESHORE LIVING, NOT JUST ABOUT LAKES OR ABOUT THEIR protection. People want to use lakes. Desire creates value, and value often creates problems. To truly acknowledge the assets the land has to offer, as Aldo Leopold, in *Game Management,* stated, "to see why it is, how it became, and the direction and velocity of the changes," is as important as seeing what the land offers today. To effectively place humans and the rest of nature together on the lakeshore, we must understand the special attraction of lakeshore living. Sigurd Olson contemplated in *Listening Point:* "Water reflects his inner needs. Its all-developing quality, its complete diffusion into the surrounding environment, the fact it is never twice quite the same and each approach to it is a new adventure, gives it a meaning all its own. Here a man can find himself and all his varied and changing moods."

Water's appeal can be traced across continents and cultures. In many lake-rich districts of Russia, for example, people seem drawn to the water. There, as in Minnesota, the Scandinavian heritage of fishing cabins seems to have grown and changed colors. Bill Holm, in *Cabins of Minnesota,* reflected beautifully on cabins, cabin life, and lakeshore living. He traced the various motivations of people to go to the cabin, making the connection with the Swedish *stuga,* or summer cabin. In Scandinavia and Russia, the whole family often moved to the cabin for the summer, with the father visiting from town as often as his job allowed. Holm also describes how Minnesotans transformed that tradition, with less time spent swimming and gardening, more time spent fishing, and the adoption and invention of new customs.

One of the primary changes in the cabin tradition was that private property became symbolically important. Although most lakes did not become private property, perceived ownership of the lake and of the lifestyle became meaningful. Owning your own spot directly on the lake, with your own pier, a private beach, and few other people around, easily captured the imagination. However, at the same time, the cabin offered a community life. Large families, extended families, and friends would share small cabins. People from neighboring cabins would visit and gather around the fire or barbeque. A private idyll was pursued in company. Holm links the Minnesota cabin tradition to the romantic tradition of retreats in nature, and even further back to Chinese lore about the pleasures of a simple life away from society. Interestingly, the idyllic views of nature are often formed in an urbanizing society, and visions of an isolated life are formed in an intricate community. It is not surprising, then, to find in a cabin many overlapping activities and many contradictory expectations, and to find lakeshore living as complex as any urban lifestyle despite all its aspirations of simplicity.

For many people, the pursuit of simplicity combines with a pursuit of timelessness, of time standing still. Sigurd Olson noted in *Reflections from the North Country* that "Time moves slowly, as it should, for it is a part of beauty that it cannot be hurried if it is to be understood. Without this easy flowing, life can become empty and hectic." Logs, fireplaces, and wooden furniture give the impression of old unchanging lifestyles. Old buildings may be moved to the lake and re-created as cabins, or old cabins may be renovated. Living in these modest secluded cabins, or emphasizing old traditions, can give the feeling of a connection to the past. There are myriad other desired lifestyles and motivations for lakeshore living, of course. For some, the lake is the networking environment for the summer, an extended golf course. Others enjoy the fishing. Some simply enjoy living in two different environments, while others desire to get away from an urban place, from other people, or from a lifestyle. Some people adopt a different persona in the cabin; others don't. Some people see nature as a scenic background for their activities, while others are indifferent; still others go to the lake wholly for the natural environment. Sigurd Olson wrote in *The Singing Wilderness,* "there are many trappers' cabins in the north and there are many mansions called cabins." Despite this variety of backgrounds, motivations, and philosophies, the goal remains consistent: a desirable lakeshore lifestyle. All kinds of people aspire to the cabin life, to a form of lakeshore living.

Many changes in the lakeshore lifestyles and development patterns were inevitable, a result of the expansion of cities and city life to lake areas. Larger developments often led to cookie-cutter patterns or to poorly planned neighborhoods in townships surrounding the lakes. Many people settling in these neighborhoods wanted to retain something of the cabin lifestyle despite the (sub)urban environment, but that proved difficult. In many cases, the lakes had become more polluted, so swimming and other activities became less pleasant or entirely impossible. The surroundings were often more agricultural than natural, and only large residential developments made it profitable to take them out of agricultural use. Isolation was harder to pretend in these places, and images of nature could not easily be projected on these environments.

The ruthless transformation of many lakes was nothing compared to the complete disregard for wetlands in much of North America. Early America strongly believed in the power and the right of humans to change the landscape. Large-scale drainage projects, particularly in the fertile American Midwest, destroyed hundreds of thousands of wetland acres. This devastation happened despite literary and artistic support by people like Thoreau, Ralph Waldo Emerson, and landscape architect Frederick Law Olmsted. They were often more fascinated by wetlands than by the lakes themselves, framing both wetlands and lakes in the same picture, as part of the environment that could bring us closer to our inner self, to life.

Barbara Hurd, in *Stirring the Mud: On Swamps, Bogs and Human Imagination,* explored the origins of the unease and the fascination toward these areas. Wetlands are places of ambiguity, neither land nor water, with changing forms, and a hardly readable landscape. They are seldom accessible, treacherous, and unpredictable. In many cultures, wetlands were places to avoid, places of witchcraft, magic, and evil. Wetlands were also difficult for many cultures to productively utilize. Some people are fascinated by this fleeting world, while others find it more difficult to understand and appreciate the value of these waterlogged areas. Hugh Prince, describing the history of wetland exploitation in the United States in *Wetlands in the American Midwest,* reinforced the image of the undervalued wetland. Early European travelers tended toward negative connotations of wetlands, not positive romantic images. Even in 1949, when Aldo Leopold proposed a new land ethic, the idea of owning a wetland to conserve its ecological value was seen as extremism. In the environmentally sensitive 1960s, farmers received subsidies to drain wet prairies. State governments eventually became more aware of the values of wetlands, and especially since the 1970s, many state laws protected and even restored wetlands. Still, many local governments, developers, and citizens see the intricate web of wetland regulations as restrictive and unnecessary.

Because of these split mentalities regarding our water resources, reinventing lakeshore living will not be easy. Aldo Leopold was clear that ecosystem protection and asset preservation (i.e., protection of rare natural features) was the ethical responsibility of the landowner. In addition, Leopold, as outlined in "The Ecological Conscience," believed that the owner "should feel some obligation as its custodian, and a community should feel some obligation to help him carry the economic cost of custodianship."

Many of the obstacles to sustainable lakeshore living originate in the planning system and system of local government. But the issue is also a cultural one; from an ecological, hydrological, and aesthetic perspective it is essential to see lakes, wetlands, and land as one and the same landscape. Yet the belief remains that one side of the equation can be isolated from the other, that lakes can be used without reference to the wetlands, or land without reference to lakes, or that wetlands can be removed entirely from the equation. Herein lies a challenge for education, for planning, for science, and for politics.

Reflecting on traditions of lakeshore living can give some grounding to analysis, planning, and design. We can revitalize and expand traditions, but ignoring them deprives us of a rich potential for combined uses and makes it politically difficult to implement our ideas. New visions can be imagined and new combinations of uses can be discovered, but in order to be effective, these new ideas need to find grounding in the existing cultural context. Site-specific design is culture-specific design.

NEIGHBORHOODS ON THE WATER

To study culture is to study communities. Planning for community should take precedence over planning for individual lots. One way to radically undermine community life, and at the same time create problems for ecology and access, is to develop the entire lakeshore. Dividing all available land into individual parcels is also detrimental to creating communities or neighborhoods. This fragmentation of the lakeshore increases the distance between homes, makes it challenging to create shared amenities, minimizes access, and ignores landscape boundaries and other ecological characteristics. Maintaining a distance between houses and the shoreline, creating a clustering of homes in a neighborhood setting, and sharing amenities, on the other hand, can save space and ecological values and improve the overall community quality.

However, one cannot do everything anywhere; understanding the context is essential. Lakes that are closer to cities, or are part of an extended metropolitan fabric, will offer special opportunities to integrate functions and special challenges with regards to the ecology. Sigurd Olson noted in "Wilderness Canoe Country," "no one likes regulation, but regulation is mandatory when large numbers of people use any area." The choices for preservation of existing ecology and creation of new nature will shift accordingly: something that might be very commonplace in northern Saskatchewan might be very special in Minneapolis, Minnesota. Its preservation in a metropolitan area might deserve more resources and effort.

Creating vibrant neighborhoods on the water without damaging the landscape beyond repair and without losing the identity of place poses a considerable challenge. Analyzing the landscape and seeing it as a coherent set of assets that can be preserved, expanded, and connected is crucial. First and foremost, we must consider water. There is no lakeshore living without water.

KEEP IT WET

Water retention is an urgent problem in much of the modern world. The disappearance of many forests and wetlands, implementation of sewer systems, and increase in the amount of pavement and roofs have resulted in water being drained much faster than in the past. Water retention locally can be enhanced by minimizing paved areas, or by creation of rain gardens, dry ditches, new wetlands, green parking lots, and small woodlands. These landscape elements can, in turn, help structure the development and should not be assigned marginal spaces. Rain gardens can be made part of the green infrastructure, and they can become attractive and useful additions to parks, ecological corridors, and greenways. Strings of rain gardens can even produce a variety of small wetlands. Water retention can go hand in hand with erosion control; wetland buffering of agricultural land, planting of slopes, and use of native plants all contribute to water retention. Lawns use excessive amounts of water during part of the year and contribute excessive amounts of runoff in the other; avoiding lawns as the universal landscaping solution also helps increase water retention. Leopold and Olson both worried about erosion, and it is clear that creating the type of shoreland development they would envision is only realistic if we solve water-retention issues.

An understanding of water circulation and water retention can also help in establishing several principles for ecological zoning of watersheds. First, upstream wet areas deserve higher priority in protection or restoration than downstream areas. Second, development in relatively clean upstream areas of relatively clean watersheds could be restricted; downstream, nutrients naturally gather, and because the diversity in downstream areas is already lowered by the natural accumulation of nutrients and sediment, developments in those areas may have less of an impact.

BUFFERS, SCREENING, AND FRAMING

The placement of the cabin or lake home on the lot is critical. As discussed previously, this decision must include intelligent and thorough site analysis. Besides designing to meet setbacks and other regulatory requirements, the building location must be respectful of place. Sigurd Olson spent considerable time thinking about where to build his Burntside Lake cabin. Rather than place the cabin on the point itself, which he thought would destroy the sanctuary of place, or at a location that could catch the full view of the lake and its islands, he chose a place that would receive the winter and early spring sun. The selected place was in the woods where the view of the lake and horizon were muted. It was a private and buffered location—a place picked with the understanding of the seasons, a place where a cabin would blend into the background of pine, mountain maple, dogwood, and rock. In *Listening Point,* Olson wrote: "While there would be no unbroken views from the windows, they would give us a hint of what could be seen by merely stepping out of doors. . . . If we could see all there was to see

BUFFERING IN TWO STEPS CONTRIBUTES TO WATER RETENTION: WATER DOES NOT DRAIN AS QUICKLY, EROSION IS PREVENTED, THE UPLAND ECOLOGY IS ENHANCED, AND ECOLOGICAL CONNECTIONS BETWEEN UPLAND AND WETLAND ARE RESTORED.

from indoors, if we became content to have the beauty around us encompassed by the four walls of the cabin, we would lose what we came to find, and that we must never do."

Buffers are a key ingredient of any design strategy of asset preservation. Buffering is simple: things that should be separated are kept at a distance. However, it is not always easy to determine what should be separated and how extensive the buffer should be. The better we understand the functioning of the landscape and the impact of our own activities, the more precise we can be in the use of buffers. In very sensitive areas, buffers must be precise in order to be effective. In calcareous fens, for example, the quality of the water and the pressure of the groundwater flow are narrowly defined, and buffers must accommodate this. Natural shoreline vegetation serves a critical role by buffering negative impacts of development. Lakes fare better when shores and slopes are planted with a diversity of vegetation. Slopes are particularly vulnerable to erosion, and erosion of sloping lakeshore is a serious problem, threatening the ecology of both upland and lake, and in some cases the physical stability of houses. Dense natural vegetation is absolutely preferred above lawns in these areas. Slopes therefore should be designated buffers for erosion.

Planting shrubs and trees can help protect the shoreline but can also help screen and frame places along the lakeshore. Screening and framing are among the most important tools used by landscape architects. Screening defines spaces and the views from these spaces; as such, it creates the experience of the place. Screening and framing are therefore design tools that

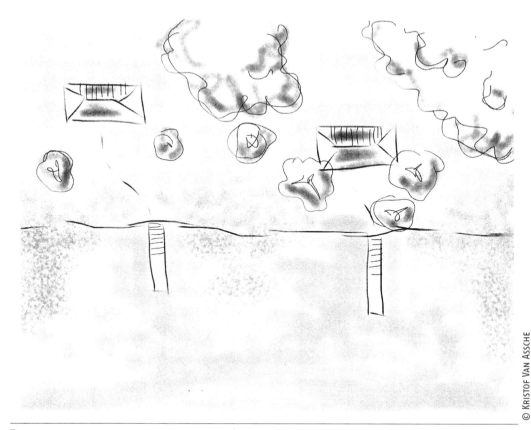

KRISTOF VAN ASSCHE

TWO WAYS OF BUFFERING THE IMPACT OF THE HOME. IN BOTH EXAMPLES, A DRY BUFFER WITH NATIVE SHRUBS AND TREES MINIMIZES EROSION AND VISUAL IMPACT. SELECTIVE PLANTING AND CUTTING STILL ALLOWS FOR VIEWS FROM THE HOMES. A SECOND BUFFER OF WETLAND VEGETATION IN THE WATER AND ON THE SHORE PROTECTS AGAINST THE WAVES, ABSORBS NUTRIENTS AND POLLUTANTS, AND BINDS THE RUNOFF SEDIMENT. IN THE RIGHT EXAMPLE, A SMALL BEACH AREA IS STILL PRESENT. LEFT, LESS VEGETATION IS USED.

enable asset preservation and asset creation. Screening and framing can shelter sensitive spots and highlight beauties that would not have been perceived otherwise. They can create a suite of perceptions and experiences while moving through a landscape. Ernest Oberholtzer's island on Rainy Lake, Minnesota, is a carefully sequenced series of visually linked spaces that are nevertheless screened and framed so cleverly that they each acquire a different character. Although the buildings and spaces on the tiny island are very close to each other, each is experienced differently, partly because of the screening. Both spaces and views can be framed; fewer but more defined views from the cabin can even increase the value of the property. Views can change between different floors in the cabin or between different elevation levels on the property. If one consistently and creatively uses principles of screening, one can create outdoor spaces that connect like outdoor rooms, each offering a different character and different views.

These principles apply not only to the individual lot, but also to the lakeshore as a whole and to the level of the lakeshore neighborhood. A series of well-screened lots gives the impression of a more natural place with more visual variation and better views. On more natural lakes, extensive screening might not be in tune with the local landscape character, turning it too much into a landscape garden. On these lakes, however, structures that are visually

© PAUL RADOMSKI

Winter view of shoreline buffering. Note the protective nearshore vegetation consisting of shrubs, grasses, and forbs. Trees frame the view, while providing screening for privacy.

unpleasant or disturb the local ecology should be buffered or screened. Screens can also serve as ecological corridors and can accompany new and old amenities. This is especially useful at the neighborhood level, where higher densities of people often create pressure on the natural environment and tensions within the new local community. Careful siting and screening can minimize the conspicuousness of high population density, close neighbors, or tourists utilizing the lakeshore.

Using landscape boundaries and natural landscape features in this effort can diminish the impression of artificiality in the screening. It is not enough to separate private, semiprivate, and public places by means of planting, water, and walls; using intelligent routing, natural landscape boundaries, and landscape elements can allow for higher densities of people while still preserving a natural character. Elevation differences greatly influence how the development and landscape are viewed. Constructions on open lots on the top of a hill will be visible from everywhere and will give the impression of high population densities and artificiality. Developments placed just below the summit, surrounded by trees and shrubs, are more likely to create the impression of more nature and fewer people.

In the design of the neighborhoods, the choice of screening materials and techniques can contribute to the neighborhood character. A neighborhood, for example, can be defined by a particular species of shrub or by a mix of species. Holly Whyte found that fences used as screening often created social barriers.

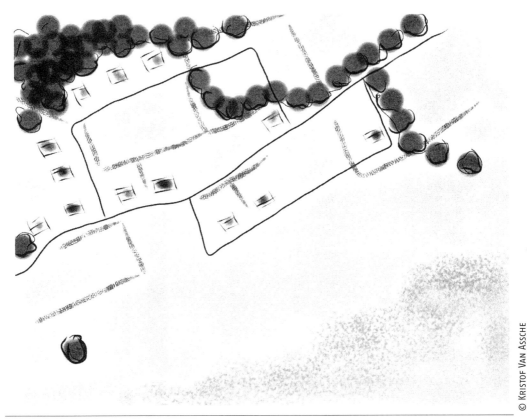

FRAMING AND SCREENING FOR VIEWS AND TO CREATE SPACES FOR DIFFERENT ACTIVITIES.

PRIVACY AND OPEN SPACES

Screening and siting define the balance of open space and closed space and determine the way this balance is perceived. For tourists and residents, the requirements might be different, and in mixed-use situations, the balance might be precarious. Tourists might want a more natural character and easy access to the lake, while residents might pay more attention to practical elements such as infrastructure and might want to monopolize lake access. A careful distinction between private, semiprivate, and public space often helps in these areas. For many lakeshore homes, the lake is the common open space. Opportunities for interaction with neighbors exist at the dock and out on the lake. Homes can have an open front garden next to the street but a backyard that is enclosed. In shoreland developments, the backyard is typically lakeshore and should be enclosed by means of native vegetation. Backyards can be separated by a walking trail, accessible to tourists but not obvious. Local residents may use these trails with some regularity, while visitors can delight in the element of surprise associated with discovery. In both cases, the backyard will be shielded from trail-goers. A more clearly indicated walking trail could connect front yards and possibly connect to large open spaces and yards of nonriparian residents. Whyte observed in "How the New Suburbia Socializes" that "people make friends with those in back of them only where some physical feature creates traffic."

Large open spaces can be in the middle of a neighborhood, as is the classic design, or they can be laid out in other patterns that enhance the visual quality of the area. A chain of open spaces, a circle of open spaces surrounding the neighborhood and extending into the woods or wetlands, or a natural pattern of open space can help define the structure of the neighborhood. The character of open space in a lakeshore neighborhood can be artificial, natural, or a combination; a neighborhood might have some artificial open space within the built-up area and some natural open spaces toward the edge. This gradual transition from nature to culture can already be found in Renaissance parks. It can improve the impression of embedding in the landscape and the impression of nature permeating the development. Creating areas with different densities, lot sizes, setbacks, and even architectural styles can reinforce this effect. In the more urban and cultural part of the neighborhood, more stone, dividing walls, and paved areas, urban or village architecture, parklike open spaces, and higher densities of people might be appropriate. Toward the edges, larger lots, greater setbacks, more natural planting and natural elements, less impervious surface, or more cabin-like architecture could be required. Higher densities could also be achieved in smaller clusters, for example, around simple natural open space with the character of grassland or clearing in the woods. Whyte saw a diversity of open spaces as a major community asset; he wrote in *Securing Open Space for Urban America:* "Open space is not the absence of something harmful; it is a public benefit in its own right, now, and should be primarily justified on this basis." Acquiring open space by easement, purchase, or through zoning is critical. In *The Last Landscape* Whyte stated that we need both large and small open spaces and

> The kind we should save first is the kind that is most useful to people—the spaces that are closest to them. If some of them are big, so much the better, but in most cases what is left are the small spaces, the irregular ones, and the maligned bits and pieces. Weaving these together is a far tougher challenge than setting aside large chunks somewhere else, but it can be done. . . . The significance of a space depends on where the space is, what it is like—range, hill, woodland, marsh—what the surroundings are, how many people use the space or see it, and when.

The ideas presented above are not ironclad rules; design of open space should be site specific and creative. The open-space network should become an integral part of the design, at the scale of site, neighborhood, community, and lake. When working with open space, it is useful to look at open space as the sum of the open spaces within built areas (both private and public) and the surrounding natural areas that are visible or directly accessible from the inhabited places. Open grasslands or wetlands directly surrounding the neighborhood should be included in the open-space design because they contribute to the experience of the space. The same applies to water, either as expanses or as channels permeating the neighborhood. These more natural open spaces not only enhance the impression of naturalness, they also reduce the amount of private open space needed for people to feel comfortable.

SAVING THE SHARDS

Increasingly, lakeshore development is lakeshore redevelopment. More and more lakes are being developed, many in ways that fly in the face of contemporary taste and sensibilities. Increasing market pressure on lakes is a double-edged sword. It can lead to irresponsible building practices, short-term strategies, and environmental pollution, but the same market pressure can also be a condition for profitable redevelopment; only when expected profits are

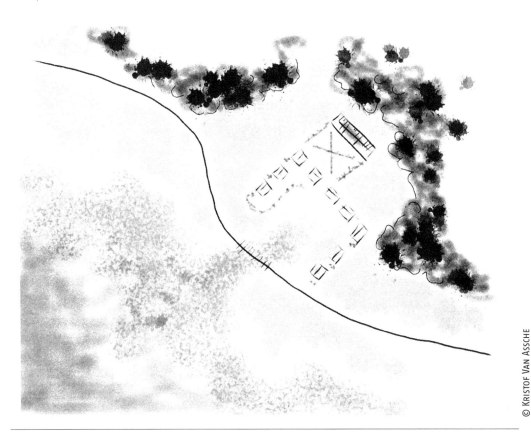

CLUSTERING OF HOMES INSPIRED BY MEDIEVAL COMMUNITIES: A SEMI-COURTYARD OPENING ONTO THE LAKE AND A CLOSED COURTYARD IN THE BACK. THE WETLAND IS ENTERING THE CULTURAL SPHERE. WITHIN THE SEMIPRIVATE SPACE OF THE COURTYARD, SOME PRIVATE GARDENS ARE DEFINED BY HEDGES. IN OTHER PLACES, WALLS MIGHT BE MORE APPROPRIATE. THE HOMES AROUND THE SMALL COURTYARD ALSO HAVE OTHER VIEWS.

high enough can powerful stakeholders be convinced to clean up older low-quality developments. These restorations are great opportunities to think more comprehensively and to work toward a design for the whole lake that embodies the principles outlined in this book. However, redevelopment offers special challenges. Ownership is usually more complex than in a new development. Undesirable structures and infrastructures may be present. Finally, there is a history of land use that already excludes a purely natural starting point. Still, renovation and redevelopment will slowly become the norm, and new strategies must be devised to accommodate these approaches. In many places, these redevelopments are taking place without a plan. Unsurprisingly, the effects on the environment and the community are often problematic. A good design should be a vision for the long term. Cities can buy lots outright or apply stricter rules for rebuilding. Zoning regulations can be sharpened or more strictly enforced, and environmental regulations can be used to push the development pattern in a different direction. The redevelopment of neighboring areas can help bridge the gap between old and new; an old edge can become a new center, a public place, or an amenity. In that manner, linear developments can gradually be turned into neighborhoods. Several lots on the lake can be purchased for public access, a negotiation for trail easements can be started, and new structures on the lake lots can be set back farther from the lake.

With financial, political, and cultural support, it is possible to reverse the trend of piano-key developments, of an entirely and uniformly parceled-up lakeshore. Holly Whyte offers straightforward ideas and principles for redevelopment. Aldo Leopold and Sigurd Olson don't offer simple and clear recipes for development as redevelopment, but one can derive design logic from their principles. In addition, they offer detailed examples of their own experiences with their cabins and the surrounding landscapes. Leopold carefully restored the beauty and the diversity of the sandy landscape around his cabin in a sustained family effort, whereas Olson focused on respecting the existing landscape and carefully and modestly inserted design elements. Leopold and Olson both reused old buildings for their cabins. Leopold made reuse of materials and objects into a guiding principle for all activities in the sphere of the cabin. In a general sense, an argument for development as redevelopment can be connected to many of the principles that Leopold and Olson developed during their lives.

Development as redevelopment may be necessary for more comprehensive management of natural areas or those areas with some remaining natural qualities. Whyte argued in *The Last Landscape,* "we must concentrate on the smaller spaces, the irregular bits and pieces, and especially those that we can connect together." Development needs to take into account the remaining natural qualities, as well as those parts of the cultural heritage that are worth remembering. Olson wrote in "The Romance of Portages": "Wilderness is a delicate adjustment, one which can be disarranged very quickly, and when in addition to its original primitive value it has bound up irretrievably with it a romantic human history, then it becomes doubly imperative that the utmost care be taken in its administration for fear of disturbing values which can never be replaced."

Leopold, Olson, and Whyte clearly understood that working with the existing landscape and cultural assets was an imperative, not just to preserve these natural and cultural qualities but also to create an identity of place. Multiple land use was unavoidable in most places. Whyte understood the need to blend different uses, whereas both Leopold and Olson had their misgivings about certain interpretations of multiple use. In "Wilderness Canoe Country" Olson said, "the slogan 'saving our wilderness through multiple use' is sound only in the proper application of the concept." He referred to some uses of the concept of multiple use—and Leopold's "wise use"—that basically opened the door to old practices of ruthless exploitation. They were aware that redevelopment should be an inclusive strategy, including natural and cultural assets, as well as including the local stakeholders in decision making. Neither Leopold nor Olson was able to predict the mixing of residential and recreational developments that are commonplace today. For them, the world of the cabin was different from the world of the house, as it still is for many lakeshore cabin owners. But, accelerated by the Internet and an aging population, this boundary between two worlds is fading. More young people are turning the lakeshore cabin into a first residence, and more of the older population is retiring to cabins. The expectations of cabin life are different than those of home life, and the pressure on the environment is growing with the conversion of recreational into residential use. Often, this pressure involves more changes to the natural landscape and more pollution. In addition, older cabins are demolished and older trails are wiped out. Identity of place is disappearing rapidly, and lacking community life is even more problematic as people spend more time on the lake. In this new context, the argument for reuse and redevelopment is even stronger. If we want to work in the spirit of Olson and Leopold and in the pragmatic discipline of Whyte, we will have to develop new design strategies to incorporate the old into the new, to combine functions, and to preserve and create identity of place.

Combining old and new, known in planning and urban design as "the use of heritage," is always a sensitive topic: some people identify with the old places and objects and others do not. Some people recognize an object or tradition as heritage but still do not value it. Designing with heritage and with older places and objects is a matter of selecting elements, structures, and places that are deemed interesting and discarding or sacrificing others. On many lakes, the legacy involves small narrow lots with homes directly on the lakes—a problematic legacy. Still, local variations appear, in narrowness, setbacks, architectural quality, technical condition, embedding in the landscape, sanitary arrangements (sewer, individual sewage systems, holding tank), and more. In other words, not all elements should be discarded immediately, even if it is technically possible to do so. Sometimes developments are retained for sentimental reasons, such as their old age, meaning to the community, or their historical human-interest stories (e.g., designed by a famous architect or lived in by a famous person).

In problematic stretches of piano-key development, complete redevelopment may include thinning out the lots (rebuilding one home on two lots) and changing the setbacks. Moving homes away from the lake and closer to the street can be a start for a new neighborhood, particularly when homes are added on the other side of the street. Varied setbacks can be used to increase privacy when there is no plan of adding houses and creating a neighborhood in the second row.

NONRIPARIAN LAKESHORE DEVELOPMENT

Shoreland development includes more than just those homes located directly on the lakeshore. Often a second and even third row of homes and cabins is present, and in old-style lakeshore developments these areas can be highly problematic. The contrast between the first (lakeshore) and second rows is often remarkable: shacks and cabins in all stages of decay can be found behind upscale homes in perfect condition. Lakeshore property is highly appreciated, and values are more stable in economic crises. The homes behind the shore might not be as well kept and may be owned by people with different aspirations and values than those onshore. In addition, many lakeshore associations only represent the people living directly on the lake. Lobbying in local politics on behalf of "the lake" will often acquire the character of a lobby for the group directly on the lake. Improvements to the overall lakeshore are rare, and access, use, and even views of the lake tend to be monopolized by the onshore group. These lifestyle differences and political maneuverings may escalate into tensions between residents.

Older developments with two or more rows of homes are most vulnerable. The offshore homes in these developments are less likely to be well maintained, are less likely to become converted into upscale homes, and have more volatile property values. The owners of these homes often have less of a voice in local politics and can profit less from the advantages of lakeshore living. In these cases, everything after the first row should be looked at closely, and the options for redevelopment should be seriously considered. In order to effectively redevelop the offshore area, one needs to revisit the causes of the initial problem cycle. In other words, some of the pleasures of lakeshore living will need to be redistributed to improve the chances of the back rows. In practice, that means new management of access to and views of the lake. A cluster of homes at a little distance from the shore will need to be supported by a clustering of lake access and public and private amenities. When considering these changes, it remains important to strive for a design with overall consistency. Sometimes small changes can dramatically improve the consistency of a plan; altering one street or structure obscuring an otherwise beautiful home can be efficient and effective. In other cases, more change is required, and the designer will have to devise a new structure for the whole development. If so, the decisions are more difficult, and more existing elements and structures will have to be discarded. Sometimes acupuncture is the cure, and sometimes the wound requires a serious surgery.

URBAN LAKESHORE REDEVELOPMENT

A special case emerges when the lakeshore development is part or becomes part of a more urban setting. In these situations, the decisions about design selection and combination of old and new will have to be informed not only by the landscape context, but also by the context of the town or city. In *The Last Landscape* Holly Whyte noted:

> Density also has an important bearing on the look and feel of a neighborhood. If it is urban it ought to be urban. Most of our redevelopment projects are too loose in fabric. They would look better, as well as being more economic, if the scale were tightened up . . . concentration is the genius of the city, its reason for being. What it needs is not less people, but more, and if this means more density we have no need to feel guilty about it. The ultimate justification for building to higher densities is not that it is more efficient in land costs, but that it can make a better city.

In the past, small lots and allowance of building connectivity via shared walls promoted higher density with a mix of retail and residential use. The diversity of development created vibrant density. It is much harder to create this sense of community with large block development, especially if starting with tall structures, as people are too removed from public lakeshore places and large blocks often lead to wide streets that deter walking. Away from sensitive areas the historic conditions of small lots and building connectivity can be replicated today.

A more urban backdrop does not force increased densities evenly along the lakeshore. Low-density enclaves, green zones, ecological transition zones, lake access, and various public and private amenities can add greatly to the quality of urban life. Public lake access and associated amenities can be designed in various styles. In general, it is advisable to incorporate lake access and lake views into the urban fabric. Holly Whyte encouraged cities to reclaim their waterfronts, and many cities now have. He saw the value of opening up the waterfront and redeveloping urban areas with lakeshore sensitivity in mind. Whyte stated in *The Last Landscape,* "There should be maximum physical access—and not just to the waterfront, but the water itself. . . . Every new waterfront project, public or private, should contribute to access." The existing pattern of streets and habitation, as well as the spread of business, can serve as inspiration.

Simply using the shore as a park does not always provide the greatest benefit. The interface between water and town is often the interaction zone, and this area creates added value for the city. If water, wetlands, and views can be drawn into the city, the spatial quality will generally be higher. Environmental risks are present, and design has to be careful. In most cases, the lake was not intended to become the facade of the town, and messy development and chaotic roads were the result. In order to turn the lake into the highlight of the city, things need to be cleaned up, and a new transportation plan for the entire city might be appropriate. Heavy traffic is a barrier to pedestrian flows, and even attractive development and amenities along the lake will be underutilized unless a friendly pedestrian connection with the rest of the city is secured. Whyte wrote, "As a user of space the pedestrian is a far better unit than a vehicle and the space now allotted him is so minimal that even modest additions could have a high leverage effect."

CHAPTER 7

Asset Creation

By now it is clear that asset preservation and asset creation are closely interlinked. An asset only remains an asset in a suitable context, and an area improved by a design can turn existing elements and structures into real assets. A feature can go unnoticed for a long time, but once made visible and given prominence in a plan, it can become a focal element in a landscape. An asset can also be created by stories; stories help make qualities visible to other people. People like Leopold, Olson, and Whyte made a big difference in this regard; they not only delivered the scientific facts or environmental education, but their writing can help us see and appreciate the world around us. Olson, in *The Lonely Land,* said: "I sat there a long time and listened to the sounds of the great marsh, the rustle of reeds and grasses, the lap of water, the far calling of loons, and finally must have dozed. When I woke, the moon was high and a path of glittering silver lay across the lake." It is in these evocations that assets are created, that an appreciation of the landscape and its interconnected elements can be born. Site design needs to be embedded in the landscape; scientific analyses can help in producing a site-specific design, but a sensitive observation of the place and a listening ear for its stories are as important. Economically, as well, it makes sense to appreciate and create uniqueness; cookie-cutter developments are more at risk to lose value. A site-specific design, despite being strongly connected to the locality, can take inspiration from other places. Developments that offer good examples of integrating hydrology or ecology, capitalizing on visual qualities, or of combining uses, whether from other culture or other landscapes, can be valuable influences.

ECOLOGICAL RESTORATION

In much of North America, people have transformed the landscape. Most of the prairie wetlands have disappeared, and the transition-zone oak prairie has become a rare landscape type. In northern Minnesota, Ontario, Wisconsin, and Michigan, much of the forest is secondary forest, regrowth after extensive logging, with different species of trees and an altered ecology in the understory. Bringing back the old landscape and ecology, with all the vegetation types and the species, is bound to be impossible, exceedingly expensive, or potentially undesirable. People are part of the ecosystem too, and agriculture and habitation have created environments with new ecological and spatial qualities.

Aldo Leopold and Sigurd Olson focused their efforts on preserving what was left of the landscape; the magnitude of the natural resource devastations at that time made restoration efforts generally inconceivable. Still, in the course of their careers, they became aware that humans and the rest of nature had to be given their place in each other's proximity. Context became ever more important for Leopold, and he believed that wise use should depend on the landscape context, as well as the cultural and economic context. Many landscapes could probably not be saved, but in other places, some restoration of ecological quality was possible. Leopold practiced restoration on the grounds around his cabin and developed plans for eco-logically enriched agricultural land, an early and modest form of ecological restoration. Olson, too, gradually became more aware of the importance of management and planning for various uses. He did not like the idea of nature restoration and believed it would be near impossible to bring back the full splendor of a forest once logged. Still, he believed that some wounds could heal and that people could help in the healing.

Now, after decades of research, we know that some ecosystems and ecotypes can be restored with reasonable effort and within a reasonable time. Certain ecosystems can be restored for ecological reasons, such as the rarity of the ecotype (e.g., calcareous fens) or the rarity of species that could return (e.g., lady's slippers). They can be restored because they form part of a larger ecosystem, such as a complex of lakes and diverse wetlands within one unspoiled watershed, whose value is primarily associated with the larger scale. Or a restored landscape can serve as a corridor between two patches of important habitat. The decision to bring back certain communities requires a thorough understanding of the landscape. Leopold favored restoring environments over stocking fish or wildlife, as he thought too much effort was wasted in placing animals in environments that were disparate from those needed for their long-term sustainability.

In addition, there can be aesthetic and practical reasons to bring back particular ecosys-tems. Trees can be reestablished for buffering and screening purposes. An old creek can be restored to generate a green corridor from woodland to lake, to serve as a basis for a recre-ational trail, or to structure or delineate a new neighborhood. Shoreline restoration can serve the purposes of erosion control and water-quality improvement. Wetlands can be restored as filters. In various locations throughout the United States and Europe so-called *helophyte filters* are used to clean water. Reed, rush, or cattail marshes established near the house or farm help clean the water, drastically reducing the size of sewage systems and the environmental risks. These marshes are often quickly established, and where the original shoreline or wetland vegetation is not easily restored, they can be suitable replacements. These marshes can also be easily used for architectural purposes: they create clean edges, separate uses or people, and can form ecological corridors for wetland species. Similarly, willow vegetation can substitute for rarer and more slow-growing types of wooded wetlands.

CLUSTERING HOMES

If we want to create carefully embedded neighborhoods in the landscape, one of the most impor-tant considerations is siting. The location of buildings in reference to each other, to the lake, and to other landscape elements will determine whether or not the development is successfully integrated into the landscape. As earlier described, Sigurd Olson was masterful in the siting of his cabin on Listening Point, away from the lake and hidden behind a large boulder. This, in combination with the routing to and from the shore, created the experience of his cabin; it is because of this siting and approach that we can still experience the place as wild and that the cabin acquires its easy role as a place of reflection on people and nature. On Oberholtzer's Mallard Island on Rainy Lake, the buildings are also sited sensitively. Each is embedded differently in the landscape and contributes to the definition of a place with a distinct character, despite being so close to each other on both sides of the granite spine of the tiny island. Oberholtzer, with his training as landscape architect, used this same sensitivity when designing a neighborhood on the Rainy Lake mainland. He did not parcel up the shore but clustered homes around an open green space, allowing many residents to overlook the lake from a distance.

The standardization of housing types and types of siting is commonplace. Many homes look alike and are located on the water in similar ways. This uniformity makes it very hard to use the landscape in a subtle manner and makes a truly site-specific design difficult. The American New Urbanism movement can serve as a source of inspiration for a new way to design neighborhoods. Andres Duany and Elizabeth Plater-Zyberk, among the most influential planners/architects of the movement, paid a great deal of attention to diverse clustering of homes. The design practice starts from the definition of a series of housing types and a number of ways to combine and cluster those types.

A traditional feature of New Urbanist developments is the refusal to make the garage the center point of the house. Garages are often detached, concealed in the backyard, and accessible through narrow, low-speed back alleys. They can also be tucked under the house. Small parking lots, small parking garages, or a cluster of garages can be designed to serve a group of homes. The decoupling of house and garage becomes more important in complex and vulnerable landscape settings, like the ones we are interested in. Parking could be located on the edges of a development, yet still respect reasonable walking distance and some form of street access to the house. This substantial decoupling of parking and home is very helpful in allowing the planner to use more different housing types and more ways to combine them. On Listening Point, one must walk to the cabin and walk even further to the lake. There is a driveway, but it stops before it invades the area visible from cabin and lake.

Another feature of New Urbanist developments is the flexible deviation from a single grid pattern for the streets. Blocks of different sizes and shapes allow for different combinations of housing types and different combinations of uses. Use of smaller block sizes (about 600 feet in length or perhaps less) in residential and mixed-use neighborhoods increases connectivity, whereas large blocks may be more advantageous elsewhere. This diversity also makes the neighborhood more attractive for pedestrians and tends to slow down the traffic. Streets can also vary considerably in width and character. In the case of shoreland development, the topography can guide the pattern and appearance of streets. A combination of paved and unpaved streets, of pedestrian trails and shared-space streets, makes it possible to optimally use the capabilities of the site while creating pleasant multifunctional spaces.

Connectivity should be given high importance. Just because the streets vary in size and speed does not mean that they must be separated consistently. Cul de sacs should be used sparingly, and even the quietest neighborhoods need to be well integrated into the pattern of streets and trails. A small street or trail surrounding the neighborhood or village can form a clear edge for the development and provide good opportunities for walking and biking and for exploration of the surrounding landscape from different angles.

Courtyards can be used effectively in individual homes, clusters of homes, or public buildings. We already mentioned the ability of courtyards to alleviate privacy concerns within higher density areas by providing a gradation of public, semipublic, and private secluded

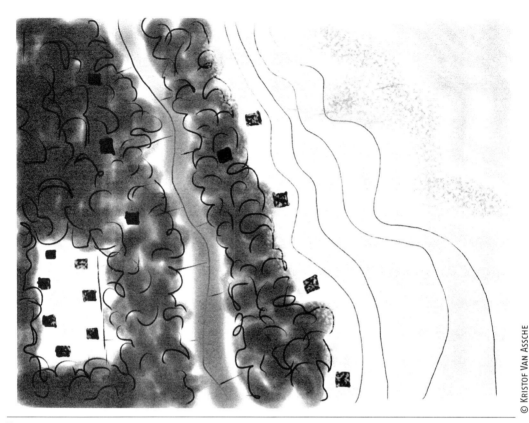

VARIATION IN CLUSTERING, AND TAKING ADVANTAGE OF LANDSCAPE POTENTIAL IN THE CREATION OF NEIGHBORHOOD CHARACTER, CAN ONLY BE ACHIEVED WHEN THE ROUTING IS NOT STANDARDIZED AND WHEN GARAGES ARE NOT DOMINANT ON THE SITE. SOME HOMES MIGHT HAVE LONGER DRIVEWAYS, OTHERS SHORTER. GARAGES CAN BE IN THE BACK OR IN THE FRONT, DETACHED OR BUILT-IN, AND IN CERTAIN SITES THEY CAN BE ABSENT. IT ALL DEPENDS ON THE DEGREE OF VISUAL AND ECOLOGICAL IMPACT ONE IS PREPARED TO ACCEPT.

spaces. However, the idea of seclusion does not have to be complete. One side of the courtyard can be left open to encompass a picturesque vista or to seamlessly integrate with a neighboring green space or cultural or natural feature. The remaining sides can consist of homes, walls, or dense planting.

When large areas of land allow the freedom to begin designing from scratch, the landscape can give guidance about the choices for housing type and clustering. Complex topographies, as are present around many lakes, can incorporate a wide variety of types. If the development is to be infill development or redevelopment, the landscape context as well as the existing homes and structure can give guidance. Local architecture, such as the Finnish vernacular architecture found in the tiny northern Minnesota town of Embarrass, may provide inspiration for other planners and designers.

When the topography is hilly, more views are available for second- and even third-row houses. The choice of clustering should take this into account. Individual homes can take advantage of good views, but homes can be arranged around the best viewpoint as well. Semi-courtyards or multilevel terraces could achieve visibility from multiple heights on the landscape. Terraces could be connected with staircases, a walking path, or a garden to offer additional views and seating in a semipublic setting.

© Kristof Van Assche

SEVERAL WAYS TO CLUSTER HOMES IN A NEIGHBORHOOD. THIS SKETCH ATTEMPTS TO EMBED THE DEVELOPMENT INTO THE NATURAL LANDSCAPE, WITH A WETLAND ENTERING A COURTYARD AND WOODLANDS EXTENDING INTO BACKYARDS, WHILE MAINTAINING A MORE GEOMETRIC LAYOUT. WITHIN A NEIGHBORHOOD, SEVERAL CLUSTERS CAN HAVE THEIR OWN CHARACTER, USE OF PUBLIC SPACE, AND EMBEDDING IN THE LANDSCAPE.

In a setting with extensive lakeshore vegetation and wetlands stretching deep into the site, a group of homes can be organized around the wetlands. The wetlands can serve as the front yard or the backyard depending on the vision of the design and the other landscape elements. Natural or artificial wetlands, ponds, or creeks can be used in various ways in clusters of homes. They can serve as center points or as edges, as dividing features or connecting elements. A creek can divide neighborhoods but can also run through a courtyard or communal park. Not every home can or should be oriented on the lake. Some homes can have a double orientation, where one side overlooks the lake, while the other side looks at a street, a forest, a public garden, or a wetland. The presence of a hill or raised area is especially conducive to a home with double orientation. In this setting, a linear arrangement of homes is imaginable. Behind the homes could be a green space that forms at the same time the front yard for a second line of homes, perched against a slope or against a forest edge.

Wooded spots can define smaller clusters of homes. Larger wooded areas can accommodate many individual homes spread around without giving a crowded impression. The edge of the woodland can also inspire a linear arrangement of homes, either using the forest as a backdrop or hiding in the forest, close to the edge. A sharp contrast between woodland and open area can be used, or the forest atmosphere can be drawn into the neighborhood by means of planting along streets and within public spaces.

© Kristof Van Assche

NOT ALL VIEWS HAVE TO BE FOCUSED ON THE LAKE. HOMES CAN HAVE SEVERAL ORIENTATIONS. THEY CAN LOOK OUT OVER THE LAKE AS WELL AS AT THE WOODLANDS IN THE BACK OR AT A PUBLIC OR SEMIPUBLIC OPEN SPACE. WITHIN ONE CLUSTER OF HOMES, A DIVERSITY OF VIEWS AND ORIENTATIONS IS VERY WELL POSSIBLE, WHILE MAINTAINING THE UNITY OF THE SITE DESIGN.

AMENITIES

Amenities can be virtually everything. In the context of lakeshore development, amenities can include those shared by the local community, such as public access to the lake, public piers, parking spaces, trails, canoe landings, and public boat storage. They can include local museums, interpretive centers, trail centers, little parks, picnic spots, public barbeque spots, landscape balconies (viewpoints to enjoy the landscape), beaches, shops, bars, and restaurants. Amenities can also include playgrounds, cultural centers, places for instruction and education, spas, public orchards, vineyards, and more. Some amenities generate money, while others cost money and sometimes become pawns in property rights battles.

The choice of amenities should be informed by the landscape analysis, but the analysis itself cannot prescribe exactly what to do. The designing process will include the opportunity to test different combinations of amenities. One will need to assess what makes sense in the landscape context, in the context of the regional economy, and in the context of the emerging plan. This sounds like a paradox, but it isn't: the emerging context of the design shapes new niches for amenities, including business, while the regional economy and the landscape have the same effect. The more the landscape, economy, and design are in sync, the better the

© PAYTON CHUNG

MODERN ROW HOUSES ALONG A STREET, WHICH ALSO SERVES AS A BIKE ROUTE, IN THE SHORELANDS OF LAKE MENDOTA, MADISON, WISCONSIN.

chances are one can create unique niches for amenities. In other words, the appropriate combination of amenities should express the character of a place while also creating it.

The features that have attracted so many people to the lakes have been severely affected by unregulated development. In a letter to the editor of the *Ely Miner* in 1948, Olson expressed that "you can't permit developments within a wilderness area and expect the wilderness to retain its drawing power." Perhaps, then, we should consider the lake ecology itself as an amenity. Value should not be placed on direct access to the lake, or even a more natural lakeshore, but on the combination of lake, wetlands, meadows, and woodland. Landscape elements that were formerly considered a waste of space, or without value, can serve to structure the development and create spatial qualities. For Olson, a rock ledge of Ely greenstone, gently sloping into the water, veined by quartz, and allowing the user to embark the canoe, or just to sit, watch, listen, was the chief amenity of Listening Point. This special place was one that would probably go unnoticed by most. Leopold and his family took great pleasure in a clump of bushes, tangled in vines, close to the cabin, and over the years took care of the shrubbery, a wonderful cultivated mess, that served as one of the main amenities of the place for them.

Rocky slopes can be considered for vineyards or berries or as a backdrop, if stable enough, for a special type of architecture. Promontories can be used for landscape balconies, be made part of local walking trails, or be used as the basis for a carefully designed neighborhood. Very gently sloping lake bottoms can be used for extensive beautification with flowering shoreline plants or as an excuse to build a long pier, ending in a platform to enjoy the views of lake and shore. Sites sheltered on three sides by slopes provide the opportunity to envision a contrasting design, a microworld invisible to most. Warmer spots, for example, limestone slopes with a southern exposure, can be used for vegetation otherwise not hardy enough. Steep slopes on the lake can be a pretext for dramatic stairs toward a pier. Lancelot Brown, an eighteenth-century English landscape architect, always saw the opportunities a local landscape offered and earned himself the nickname Capability Brown.

COMBINING USES

Mixed-use can mean many things. In the case of lakeshore living, it should mean a subtle and site-specific mix of uses. "Nature" as such does not say much, nor does "residential" or many of the other terms often found on zoning maps. Olson, Leopold, and Oberholtzer were all in favor of some form of zoning for multiple land use but had their differences. Olson supported development on the edges of protected areas, but not in them; he was opposed to Oberholtzer's idea to allow logging in the Quetico-Superior country behind a strip of untouched forest along the shores. Leopold believed less than Olson in the power of governments to control land use; he believed that in the long run, education was the only strategy that might work. Holly Whyte spoke often of the need for mixed-use in many places, especially in city centers. Discussing the need to incorporate the gathering areas of ancient Greece into today's cities, Whyte stated in *City: Rediscovering the Center:* "The agora at its height would be a good guide to what is right. Its characteristics were centrality, concentration, and mixture, and these are the characteristics of the centers that work best today."

Combining uses brings problems but is necessary for a landscape in which people and the rest of nature can coexist. A design can help in finding and shaping appropriate combinations of uses. It is important to learn which uses can be combined, how others can be buffered, and how those buffers, in turn, can be used in multiple ways. People perceive spaces, not zoning categories, and they experience activities, many of which can be combined in one space. In the case of shoreland development, perhaps the most obvious conflict in functions is between intensive agriculture and the surrounding ecosystems. Intensive agriculture depletes the soil and water resources, changes the patterns of surface water and the groundwater flows, and impoverishes the ecology through excessive use of fertilizers, pesticides, and herbicides. Lakes and wetlands surrounded by intensive agriculture face many threats. However, agriculture may not be a monolith, and several types of agriculture can be combined with forms of recreational use, residential functions, and even ecological functions. Orchards and vineyards, for example, offer recreational opportunities, a beautiful backdrop for homes, and ecological value by themselves. Dairy farming, given reasonable manure management and wide, vegetated swales in wet areas, can offer positive contributions. The swales can act as ecological corridors, and trails can be laid out beside them to provide recreational opportunities. It is helpful to see every space as potentially multifunctional. What could it be used for? How could landscape features and elements be used in several ways?

When considering combinations of uses, one can refer back to the analysis of the landscape. The hydrological system can present ordering principles for uses. Hydrology and soil, together with existing ecology, can help suggest the potential ecology, the existing and potential ecological connections, and the potential sensitivities and combinations of uses. Knowledge of plant communities, their value and vulnerability, and their dependence on certain soil and water conditions can greatly improve the quality of the landscape analysis as well as enable the creation of a site design that preserves and creates ecological quality, while creating qualities for other uses. In *Runes of the North*, Sigurd Olson describes his quest for the perfect well. His detailed knowledge of soil and water currents and educated guesses about seepage from the lake led to minimal intervention with good results. The tiny well, in turn, was the basis of a shade garden under the big conifers on Listening Point.

A MIXED-USE DOWNTOWN AREA WITH RETAIL AND RESIDENCES.

.

CHAPTER 8

Connecting People and Things

NEIGHBORHOODS THAT FOSTER COMMUNITY AND RESPECT THE ENVIRONMENT ARE more than collections of people. Designing a community to connect people is always tricky. Architects and planners have often overestimated the power of design to work on social structure, social cohesion, and equality. Many of the most problematic urban neighborhoods were designed with the best intentions but without serious thought to the people who would live there and their needs. Research and experience have taught us that certain types of design, such as lakeshore piano-key developments with no public access and no shared amenities, will probably not lead to a flourishing community life. Holly Whyte, whose studies clearly demonstrated that physical factors influence community, noted the shortcomings of excessive cohesion that can be self-destructive to the individual. In *The Organization Man* he wrote: "The comparison of physical layout and neighborliness will show that it is possible deliberately to plan a layout which will produce a close-knit social group, but it also will show that there is much more of a price to be paid for this kind of neighborliness than is generally imagined."

Community needs to develop organically. When people in a place are used to a community life, to shared activities, to neighborliness, and to shared values, then community life is easier to perpetuate. Creating this lifestyle is harder, and creating it by design is much harder. The New Urbanism movement, despite designs based on widespread principles, has not been completely successful in their goal of returning to the social graces of a traditional neighborhood. Their neighborhoods can be pleasant living environments, but social cohesion is often lacking; where it is found, it is often because a very similar group of people sharing very similar values bought the properties. Whyte, in *The Organization Man,* noted that the initial residents of a neighborhood could have an important effect on the early spirit of a community by setting traditions. The realization of an actual community life, though outside the scope of this book, can be jumpstarted by the development of a neighborhood with strong place identity. The clustering of homes in neighborhoods, the creation of functional and pleasant outdoor spaces, the implementation of shared amenities, increased connectivity and access, and a development based on sustainability will all be valuable in creating a community environment. A pleasant environment can bring in many different people. Though many will get along, some will not, and some will not like the whole idea of a community life on the lake. This forms the basis of two more principles: first, try not to please everyone with one development, and second, create diverse living environments within a community.

Even if people like the general principles behind a plan, they might not like the specific form, or they might just not like a specific group of people that happens to live there already. One worldwide principle regarding neighborhoods is that some upscale houses, well sited, can contribute to the beauty of the place and to the financing of the project. This allows those residents to fulfill the desire to be part of the community yet to be left alone. Vulnerable sites should be avoided, but places like promontories or peninsulas can be suitable. The homes may not necessarily be hidden, but the architectural quality should be high, and requirements for size, height, and other features can be specified. Even within a small community one can distinguish a few types of architecture and a few types of clustering. This might include different lot sizes, different placement on the lot, different visibilities front and back, different ways of screening and framing the views, different access to amenities, and different distance to the edge.

It is essential in this effort to create a community to think in terms of *spaces,* not of lots. A collection of lots will not achieve a neighborhood feel. The actual space, the distribution of volumes and views, will shape the experience much more. *Lot lines are not perceived unless marked by volumes or masses.* For example, the size and placement of homes influence our perception of space. A reasonable design will be ruined when homes are placed without thought at any location within a lot. Using natural boundaries for lot lines will help make a seamless transition between two homes. Defining exactly where particular homes can go is also important. Lots need to be diversified in size, and in a good plan, the largest lot will not necessarily fetch the highest price. Higher densities of homes, that is, groups of homes looking more like a neighborhood, will usually be found at a short distance from the lake. The actual lakeshore is often vulnerable and rare and may prohibit extensive development. This layout has been utilized for thousands of years; European villages in the Middle Ages were often located with access to water but at a distance from swampy or flood-prone places and at a distance from the woods higher up the hills.

If there is no central place for community activities, the creation of such a location will provide a focal point for amenities and businesses. When such a central place exists, a growing lakeshore community can help expand it, add amenities, try to attract specialized business, and integrate it into a network of public spaces, trails, and natural open spaces. In addition, children's play areas should receive special consideration beyond the ball fields and swing sets. As described in "How the New Suburbia Socializes," Holly Whyte found that children were critical in setting the web of friendship for a neighborhood, and they often did not use structured playgrounds. Instead, "children have a way of playing where they feel like playing, their congregating areas have not turned out to be where elders planned them to be." However, not all community activities, whether children- or adult-focused, should be located in such a center, and secondary gathering areas can be dispersed throughout the development. Holly Whyte observed that adjoining porches in a court or shared driveways in a dense single-family neighborhood were important factors in developing friendship and neighborliness. A network of walking trails connecting small open spaces and connecting amenities can take over the same function. The network becomes the center in that case. Intensive use is a prerequisite in that model, and it is important to assess how people can be encouraged to use the trails and to move around in the network.

The regional context is important when structuring a development. How remote is the location from larger urban areas or from one larger city? Will the development become an extension of an existing town? Understanding the context will help in deciding what kind of amenities to include, and that in turn can help in assessing what kind of community life can be envisioned. Playgrounds make less sense in a retiree community, unless the locally living grandchildren are regular visitors. Families will attract different businesses than retirees.

Income levels are important, but so are the lifestyles people aspire to. The concept of lifestyle is important when discussing identity of place. People come to places because of certain images of the place, but also because of images of the activities they can do there, of the kind of life they can have. This was true thousands of years ago, when the Romans built their country houses to be closer to the virtues of nature, to lead a simpler yet more intellectual life. And it is true now for many people who build a cabin on a lake up north. The mere word "cabin," used to refer to houses that would be labeled "villa" or "mansion" in the rest of the world, already points at a certain image of place and lifestyle, a longing for closeness to nature. Good design is critical to create the impression of unspoiled outdoor life for more people in a smaller area.

The regional context should inform but not dictate designing decisions. Larger urban areas often include pockets of a more rural and natural aspect, and those can be valued precisely because they differ so strongly from their surroundings. The opposite might also be true. The city of Ely, in northern Minnesota, is an example. Located near the pristine Boundary Waters Canoe Area Wilderness (BWCA), it is remote and surrounded by forest, lakes, and wetlands. Yet the place has a remarkably urbane feel and is attractive for intellectual and artistic travelers and residents alike. A unique combination of lifestyles is possible, although some lifestyles fit better together than others. Hunting and nature preservation can go hand in hand, whereas intensive agriculture and nature preservation are so strongly different that combining the two will not be successful. People looking for a calm intellectual retreat might not appreciate sharing their time with boisterous teenagers on a campground. General labels like "outdoor living" have many different shades, and people might be disaffected by a slightly different shade. Creating a good balance between proximity and distance, between community life and privacy, is not an easy task. Lifestyle concepts can help in creating place identity and connecting this identity to a community life, but difficult choices will need to be made.

The political organization of a place will also play a role in shaping its character. When the lake residents are year-round residents, and the lakeshore community is a substantial part of the township population, then the political will of the lakeshore community will be important. If the community is a loose collection of summer cabins, those summer residents will have less of a say in the decisions such as choice and placement of amenities. When year-round retirees reside in the community, even a small group of activists with lots of time and strong interests can influence local politics. Finally, lakeshore associations can strongly influence the political leanings of a community. Although sometimes these organizations are predominantly social clubs, other times they are well organized and political and can strongly support or oppose public amenities, environmental regulation, and more. This political and institutional setting will color the local decision-making in practice. Developing a design that garners the support of the entire range of decision makers is essential.

WATER ROUTES, STREETS, TRAILS, AND SHORTCUT PATHS

Routing contributes heavily to the experience of a place and the creation of community. Routing, as it defines the quality of approach and presents and frames objects and sites, is critical in creating place identity. For Leopold and Olson, much of northern North America was canoe country, and Olson was obsessed with the vast expanses that could be unlocked thanks to the canoe. But the canoe was much more than that; for Olson, it symbolized a way of life, a slow and thoughtful uncovering of places. Paddling gave a privileged approach to observation and experience of the riches of the lake country. Much of the beauty and variation of these landscapes can best be savored from a canoe; when combined with the experience of portaging a canoe over hidden and barely recognizable trails, the experience cannot be surpassed. The canoe was so intricately part of the experience of lake wilderness that it became synonymous with place identity: parts of the northern United States and southern Canada were designated as *canoe country*, and in some of these areas (such as the BWCA in Minnesota) this place identity became institutionalized. The canoe links up small scale and large scale effortlessly, and portages serve as markers of long histories of intense experience of the landscape. Sigurd Olson wrote in "The Romance of Portages": "And hardened tough a woodsman may be and loath to admit it, still deep down his rough exterior is an appreciation of setting and atmosphere that is second to none, and old portages to him are sacred."

Both portages and jumping-off points (those transition places on the edge of civilization, where adventure and nature begin) can be reinterpreted in lakeshore living. Jumping-off points can be brought to the backyard, and they can be incorporated in plans and designs. Just as the call of the loon recalled the northern lake-rich districts for Leopold and Olson, wilderness ideas and images of nature in faraway places can be triggered by occurrences in more cultural landscapes, even in mundane places. Unexpected vistas, approaches, sounds, or rhythms of movement can all contribute to the transformation of a place into a jumping-off point. Design cannot force such a reinterpretation of the landscape, but it can create the conditions.

This general philosophy of jumping-off points and portages can have very precise local implications. Slow movement through the landscape should be encouraged and enabled; the quality and variety of views should be managed very carefully. In *Runes of the North* Olson reflected: "By the time we were on the deck, the Radium Gilbert was moving down the channel and heading up the coast. What a change from riding in frail canoes! The headlands of the Bay, while still overpowering, had lost something of their menace, for now we were warm and safe, and in full control once more. Strange, I thought, what the consciousness of steel and power can do to a man's perspective."

Views of the water and from the water should be included in this management of approach and perspective. In larger areas, the experience should not be entirely created by the planner, but it can be enhanced. Detailed observation of the variety of assets on-site will help in identifying options for routing that might create wilderness experiences. Portages might be represented by places where movement is possible but not easy; these semiconnections might allow motivated people to move through but not others. In an undeveloped place with remaining natural qualities, where the houses and amenities can be clustered as desired, the approaches to these areas can be planned more freely as well. In case of redevelopment, the routing becomes relatively more important because the other factors can be less easily manipulated.

When dealing with shoreland developments, and especially when moving away from the actual lakeshore, it is advisable to apply some form of separation of the traffic flows. In some places, hikers and cars can use the same street, but this mixing of uses will not always be appropriate. In general one road should lead to the actual development without using too much space and without scarring the landscape too much. Some parking space can be provided within the development, but additional parking space with a green frame can be implemented on the edges of the neighborhood. Given the relatively small size of most of the developments, walking will be a realistic option, and minimizing the parking space in the actual development can increase densities while preserving a rustic character. The parking spots in the neighborhood can be partly on the street or back alleys and partly in garages.

Homes in the neighborhoods can be placed closer to the street; suggested setbacks range from twenty to forty feet. These changes minimize driveway lengths and reduce overall surface imperviousness. Porous pavers, narrower driveways, or shared driveways can sharply reduce the typical four hundred to eight hundred square feet of impervious cover created by each driveway.

Street layout is the foundation of land use design. Their patterns and standards are critical. Streets should only be as wide as necessary to meet traffic demands and to provide emergency vehicle access. Wide streets create large amounts of impervious surface cover, which leads to runoff and reduced water quality. Wide streets have faster traffic speeds than narrow streets, making wide streets less safe for people. Wide residential streets are often created by inappropriate application of high-volume road-design criteria and the perception that on-street parking is needed on both sides of the street. Communities should consider designing better streets. Street standards should emphasize pedestrian safety, and this means narrowing the street and slowing the traffic. First, the use of queuing streets (or single-loading streets) is one technique for reducing street width. Queuing streets have one designated travel lane and two queuing lanes that can be used for travel or parking. Second, low-density residential street widths could be twenty-two feet or less. Bike and walking trails can be an integral part of the design and can contribute to the sharing of space and the more leisurely experience of place. A parkway system within the shoreland can add to the sense of place. Streets can go around the lake at a safe or development appropriate distance from shore. Finally, street networks should be designed to be expandable, with the pattern interrupted by lakes and protected sensitive areas. The use of a grid pattern in flat upland areas is preferred, as it increases walkability and economic value.

Cul de sacs in general should be used very sparingly, and only after they emerge as the most appropriate solution for a site *after* site analysis and design. Many communities require the ends of cul de sacs to be fifty to sixty feet in radius, creating large circles of little-used impervious cover. One option is to reduce the radius to forty-five feet or less. In addition, vegetated swales can be used as an alternative to curbs and gutters in these designs.

Whyte found that "street width and traffic determine whether or not people make friends across the street." The number of traffic signs should be minimal, in keeping with the rustic character, and also to make drivers more careful—this is part of the shared-space concept. Sidewalks are often unnecessary under these conditions. Within the neighborhood, certain paths or trails can be restricted to pedestrians and cyclists. Some of these might be useful primarily for the residents, some might create interesting loops in and around the development, offering varied views, and others might link up with longer trails around lakes, wetlands, or wooded areas. It is reasonable to connect walking trails with normal streets if these streets are calm enough.

When designing walking and cycling trails, it is important to keep in mind that walking and cycling address different scales. However, connectivity is the essential principle; a network of connected trails, even if they are of different types, is more valuable than a collection of loose ends.

Where it is possible to build a trail along a stretch of shoreline, several options are imaginable. When the land is public or semipublic, the trail might follow natural edges and meander

© Kristof Van Assche

TRAILS CAN BE BUILT ALONG THE SHORELINE AND PARTLY IN THE WATER. A BOG WALK OR BOARDWALK CAN MAKE CERTAIN PLACES AND VIEWS ACCESSIBLE WHILE CONTAINING THE FLOW OF PEOPLE TO A NARROW STRIP. IN THIS SKETCH, THE PATH MEANDERS THROUGH DIFFERENT TYPES OF VEGETATION, AND A PLATFORM IS ADDED FOR SEATING.

along the shoreline. If a narrow strip of land near private residences is made available for the path, care needs to be taken to minimize impact for the neighbors. Walking and hiking trails should be made interesting: it is possible in most cases to give visitors a variety of views, scenery, and landscape types without affecting the most vulnerable ecologies and without imposing upon the neighboring property owners.

Subtle strategies can work: one can observe a rare feature from a distance, but the distance will likely be accepted when getting closer seems "naturally" hard. Thick or thorny bushes can be planted, or water and wetlands can create barriers. A sensitive but attractive place can be made accessible, but in a very restricted manner. Bog walks, boardwalks, narrow trails leading to overviews, landscape balconies, and very modest picnic areas invite the exploration of a place while keeping the disturbance to a minimum. Sensitive lakeshores can be made accessible for the public and private owners by clustering private docks here, public access there, and keeping other stretches natural. When discussing routing, it should never be forgotten that a street, a trail, or a path is always more than a device to get from A to B; it is always a place in itself, and it is always a way to frame the experience of the area. This means that locating routes and locating houses, amenities, and open spaces cannot be separated: they are part of one and the same design process. Choices in the location of homes should influence routing decisions and vice versa. Changes in one aspect, such as open space, should have consequences for the other aspects, such as routing. Planning and design is an iterative process, in which one gradually tests options and refines decisions. A series of trials will produce a more site-specific and consistent plan. Aldo Leopold was correct in stating in *A Sand County Almanac,* "to build a road is so much simpler than to think of what the country really needs." The route of a street or trail is critical in shaping ecological consequences and protecting the integrity of place.

Special mention should be made of entrances: entrances are symbolic points. When the entrance to a neighborhood is marked, the place is identified as a neighborhood. Place identity rests on difference. Marking a boundary or the crossing of a boundary, such as placing a clearly recognizable bridge over a creek, can create the experience of entering into a defined neighborhood. In old towns, such symbolic entry points, which often coincided with landscape boundaries, were much more important, and they contributed greatly to an identity of place now often gone.

Routing can also capitalize on special landscape features by turning them into landmarks, and it can attempt to create local landmarks by adding elements in the neighborhoods. Creation of a new place identity is more likely to be successful when the specific qualities of the place are recognized and utilized. This is what the English landscape designers in the eighteenth century called the quest for the genius loci: the genius, or the spirit of the place, captured and transformed in a plan. Using natural elements, as well as old and new artificial elements, to create landmarks should fit in this quest for a new local identity, this quest for genius loci.

ECOLOGICAL CONNECTIONS AND NETWORKS

Nature is a network of connections, between animals and plants, but also between places. Ecological networks allow species to move, to expand, and to adapt. Leopold, Olson, and Whyte were masters of observing ecological connections. In one study, Leopold reconstructed the world of quails by observing their movements, feeding spots, and nesting areas; in other words, he observed their networks. This allowed him to draw astute conclusions on the management and design of cultural landscapes that would allow them to thrive. For Olson, the discovery of the intricate network of connections between living beings was a revelatory insight, one that would shape his perspective for years to come. Designs can identify, protect, and even restore ecological networks. However, as Whyte observed in *The Last Landscape:* "It is pointless, and cheeky, to superimpose an abstract, man-made design on a region as though the canvas were blank. It isn't. Somebody has been there already. Thousands of years of rain and wind and tides have laid down a design. Here is our form and order. It is inherent in the land itself—in the pattern of the soil, the slopes, and woods, above all in the patterns of streams and rivers."

The connectedness of nature is visible in multiple aspects, from food webs to animal and plant movements. It is essential, however, to be aware that not every animal moves in the same networks and that not every network can be combined easily with other uses. Migratory birds require suitable patches of habitat in places far between their beginning and ending points, while snails require connections between suitable patches of habitat to reproduce. Bats use hedges and tree lines to assist in spatial orientation as they move across a landscape, and butterflies require a combination of flowery meadows and shrubs to complete their lifecycle. The diversity of life makes it impossible to plan an ecological network consisting of patches and connections that will work for every type of plant and animal. Once more, only careful observation and precise design will help in finding more combinations of uses and more combinations of ecological networks and routing. A thorough landscape analysis that identifies the plants and animals currently present, as well as those potentially present given the local hydrology and conditions, is crucial. If a certain type of woodland, grassland, or wetland can be restored, along with its associated species, the chances of success are greatest if one understands the ecological network those species require or could use.

Routing and ecological networks can be combined in certain circumstances. Low-impact human activities, such as hiking and cycling, can be combined with hedges or strips of wetland, a creek, or patches of woodland that serve as stepping stones for animals and plants. If a highway is present, its function as an ecological corridor will be harder to plan but is not impossible; even in these settings, a wide, vegetated buffer can guide the movement of some species. In general, the higher the density of human habitation, the higher the impact of human activities in an area, and the more dispersed the suitable patches of natural habitat are, the more important the ecological connections are and the stronger the argument for designing such connections. In urban environments, lakes and wetlands have a relatively high value, and ecological connections can be critical for the survival and reintroduction of species valued by the urban residents. In more recreational settings, the presence of natural areas and of certain species will be valued highly by visitors. In more remote areas with high but unrealized ecological potential, corridors and connections could be improved for the purpose of climate change future-proofing.

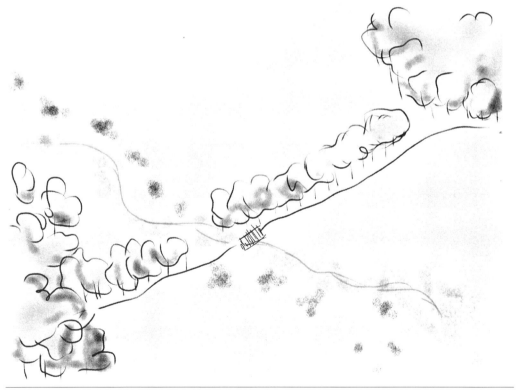

© Kristof Van Assche

CONNECTION THROUGH LINEAR ELEMENTS. A TRAIL CAN BE LINKED WITH A CREEK AND ACCOMPANYING WETLAND COR-RIDOR, WHILE ANOTHER TRAIL OR A STREET CAN BE SIDED BY TREES OR SHRUBS. THE CROSSING OF TREE LINE AND CREEK CAN CONSTITUTE AN INTERSECTION FOR PEOPLE AND ANIMALS ALIKE. SOME ANIMALS WILL STICK TO THE TREE CORRIDOR OR THE CREEK CORRIDOR, WHILE OTHER SPECIES CAN USE BOTH OF THEM AND CHANGE DIRECTIONS.

Landscape boundaries can be good places to work on ecological corridors and good places to combine hiking paths or other infrastructure with ecological connections. Edges of forest and meadow, wetland and upland, wetland and water, or valley and slope offer variations in elevation, vegetation, or wetness. These transitional areas are likely to be rich in species and are often used by animals as routes or ecological connections. Designers can capitalize on this by restoring old landscape boundaries, for example by planting a mantle of vegetation around a woodland, or by restoring a transition zone with shrubs from wetland to woodland. They can also use landscape boundaries to guide the routing; following or crosscutting landscape boundaries will offer variations in movement and will enhance the experience of the landscape.

Finally, connections and networks in urban shorelands can be enhanced with the removal of roads and highways. Streets are preferable to roads near lakes. Roads separate us from our lakes. Roads also separate people. By removing roads we can mend neighborhoods together. Increasingly, the high costs of infrastructure replacement force us to rethink the merits of the existence of some roads and highways. Several cities have already faced those economic constraints and have removed highways and converted the space to boulevards or new public open spaces. These cities have found that the car traffic adapted, property values increased, and the new public spaces were heavily used.

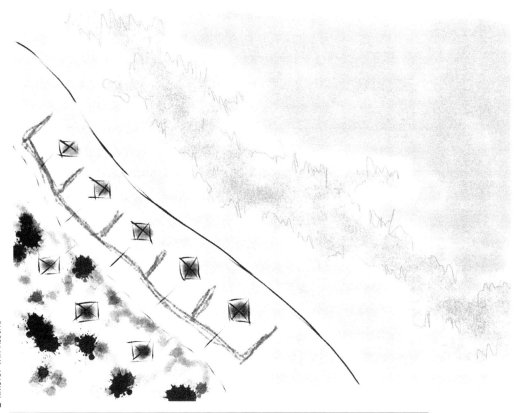

ROUTING CREATING DIFFERENT SPHERES BACK AND FRONT: STREET IN FRONT, TRAIL IN THE BACK.

A STREET IS NOT A ROAD. THIS IS A STREET, NOT A ROAD OR ITS HYBRID, A STROAD. TRAFFIC MOVES SLOWLY ON A STREET, AS A STREET IS DESIGNED FOR PEOPLE. TRAFFIC ON A ROAD MOVES QUICKLY, AND IT IS DESIGNED FOR VEHICLES.

PART FOUR

Making Good Things Happen Around Us

MAKING GOOD THINGS HAPPEN AROUND US REQUIRES UNDERSTANDING OF ECOSYS-tems, human behavior, government, and socioeconomic systems. If we want to work toward the vision of healthy, sustainable lakeshore living, then our ideas must be translated into governance and planning institutions. And these will only function if citizens trust them, see their value, and are willing to cooperate. Education plays a critical role in changing a person or a system, but for many issues education alone is not enough. We assume that if someone just understood an issue better, then they would change their ways or behavior. We assume that if we just passed a law then our problems would be solved. In practice, we will need a mix of education, communication, laws, and plans, as well as an openness to diverge from some of our specific goals in order to maximize public support for the effort as a whole. The first step is the widespread realization that present development practices are not sustainable and that design is a significant part of the solution. Once this is accepted as a premise, on both sides of the political spectrum, in the legal system, and by citizens, a new space for negotiation opens up. In the next chapter, then, we will elaborate why governance is important and which policies and tools in the North American tradition are important for us. After that, we will discuss how those tools can be used to change our lakeshore living and to implement some of the ideas presented earlier.

CHAPTER 9

Culture and Governance

THE LOON HAS NO POSSESSIONS. IT OCCUPIES THE LAKE WITHOUT REGARD FOR WHO owns the shore or how opulent the neighborhood. The loon is looking to make a good life, and it seeks a connection with place and a territory to raise a family. This bird has evolved to be a great swimmer and fisher of northern lakes and southern coastal waters. The loon knows its neighbors. Nonbreeding loons congregate on the lake in the summer, and during migration their sense of community expands into large flocks floating on our largest lakes. On their wintering waters they gather together at night for peace and security. Lakes have made the loon. What of us?

Does the love of the land and lake begin in childhood? Does the love of place shape your values and land ethics? Aldo Leopold, Sigurd Olson, and William Whyte each had a childhood filled with intensive and extensive experiences with nature and landscapes. As adults, each dedicated his life to the protection of natural places. Does a place change your perceptions of the world? Richard Platt, geographer, said that William Whyte was a diehard environmental determinist, in that Whyte "believed that the design of shared spaces greatly affects the interaction of people who encounter one another in those spaces and their resulting sense of well-being or discomfort." It can be said that people make places. We also believe that *place makes people.*

An ecosystem is variable and dynamic in space and time. In fact, it is nearly impossible to delineate its margins at a human scale; it is too broad and complex to fully understand and mentally grasp. Our perception of time is often constrained to the near present, and we are biased in our recall of the past. As terrestrial beings, we generally have a poor sense of underwater places. Ecologists attempt to fill these gaps with studies of plants, animals, communities, and dependences on earth and water. Numerous scientific journals are dedicated to collecting and sharing the insights from ecologists. Through experiments and mathematical models, ecologists have advanced our understanding of ecosystems considerably in the last sixty years. However, our ability to predict or to foresee the consequences of our actions on the environment remains, at times, insufficient to provide timely and convincing evidence for policy makers given the demands of an ever-growing eco-illiterate population. Ecosystems are chaotic in nature and sometimes too diverse for us to accurately convert to mathematical equations (and when we do attempt to simplify such things as population dynamics those same algorithms may show chaotic behavior inconsistent with reality). Ecologists will continue to advance our understanding of our world, but philosophers and poets may do a better job at capturing the nature of nature for citizens. The poetry of place can speak to the aesthetic, social, and ecological value of space or location, and the poet can speak to the mystery and importance of place. Poems can capture our passion for place in a language that we can often truly understand and appreciate.

WHAT DEFINES A PLACE?

Scientists may define a place by referring to a type of experimental unit or system. These units may be lakes, plots of land, or neighborhoods. Scientists may also define a place by its watershed or location within a watershed. This watercentric view of place is beneficial in that it provides some monitoring and land-use management benefits. However, this definition of place is inconsistent with political boundaries and, perhaps more importantly, how most people generally view a landscape or location.

If place makes people, then what is a place? Bill Siverly and Michael McDowell, editors of the poetry journal *Windfall,* define place as a specific geographical space along with its associated plants, animals, geology, history, and people who have lived to understand it or worked it to create a way of life different from that of other locations. They define poetry of place as writing about the spirit of a place, not just its location. They note that "a poem's abstractions, philosophy, and generalizations become real through a vivid evocation of the natural world: the poem's ideas are made substantial, and our subsequent experiencing of the natural world becomes tinged with the poem's ideas." They and others witness that our culture tends to homogenize the human-made elements of place, and often we homogenize the plant and animal communities with our planned and accidental transplantation of flora and fauna.

As we become more urbanized and experience less intimate time with nature, we risk losing our understanding of place. Where and when does the water run in the tree-lined valley floor or in the thick forest hills? Which birds inhabit the shoreline of my lake? What species of frogs can be heard in the early spring? Where does the dragonfly come from? How old are the bass I can see from my dock? The answers to these questions may elude us, even though we know the details of where to find the best deals on clothes or the best restaurants in our region. Urbanization does not mean that we should ignore nature, as the history, hydrology, and biology of urban places often remain dynamic and rich.

A PLACED PERSON

Wallace Stegner, in *Where the Bluebird Sings to the Lemonade Springs,* quoted his former student Wendell Berry as saying that "if you don't know where you are, you don't know who you are." He believed that a space became a place only when it was captured by literature and people's attention. Stegner also described the difference between a displaced person and a placed person. The displaced person is transient and not settled in a place. Stegner, speaking of his own placelessness in youth, noted that a displaced person knew something about many locations but was culturally lacking; he was a "discarder or transplanter, not a builder or conserver." Leopold, Olson, and Whyte, like Thoreau before them, came to perceive the uniqueness of place and set themselves into place by venturing into the natural and human-altered world. They spent considerable time walking, hiking, or paddling across the landscape and studying the plants, wildlife, people, and spaces. They discovered the rare, the typical, and the interesting and observed with keen intelligence and awareness. Stegner wrote: "Some are born in their place, some find it, some realize after long searching that the place they left is the one they have been searching for. But whatever their relation to it, it is made a place only by slow accrual, like a coral reef. . . . Neither the country nor society we built out of it can be healthy until we stop raiding and running, and learn to be quiet part of the time, and acquire the sense not of ownership but of belonging."

Geographer Edward Relph proposed that the more strongly an environment generates a sense of belonging, the more strongly that environment becomes a place. He labeled this phenomenon as "insideness." Likewise, the more inside a place the person feels the stronger will be his or her identity with that place. Relph used the term "outsideness" (rather than displaced person) to describe people who feel some sort of separation from a place (e.g., homesickness in a new place or returning to a place after a prolonged absence). We can experience different levels of insideness and outsideness for certain locations, and these experiences influence our feelings and perceived meanings of place.

To attain full, rich lakeshore living we need to become placed persons. We need to develop a better understanding of our sense of place, and we need to deliberately ground ourselves in new places. Relph advocated an authentic exploration of place—"a direct and genuine experience of the entire complex of the identity of places—not mediated and distorted through a series of quite arbitrary social and intellectual fashions about how that experience should be, nor following stereotyped conventions." Unconventionally, Sigurd Olson would paddle out into the open water of a lake and lie on his back looking at the stars while the canoe moved with the water. He took his children out in the middle of the winter night to hear the wolves and see the skies. Slowly, he and his family became grounded in the place, became part of it.

Knowledge plays a role in grounding and appreciation. For many people, getting to know a place or its creatures can deepen their appreciation of it. Our understanding of place often starts with knowing some of the common native plants and animals. The accumulation of information on organism life histories and how they interact with other organisms and their environment grounds us further. This knowledge can be a sign of being acquainted intimately with a place. For others, the experiences, activities, and memories associated with a place are most important. Some feel strongly attached to a place by simply observing, absorbing, or creating narratives revolving around it. And still others head out into the museum of natural art, observing the beauty of fractals in nature or listening to nature's music. Small shards of nature still provide great opportunities to connect with the local flora and fauna. Many people know several of these routes to grounding but do not travel them often.

The reasons to become a placed person are numerous. Scott Russell Sanders, in *Staying Put,* wrote that "many of the worst abuses—of land, forests, animals, and communities—have been carried out by 'people who root themselves in ideas rather than in places.'" A placed person may do less damage to the neighborhood. Naturalist Robert Michael Pyle, in *The Thunder Tree,* referred to an extinction of experience—the loss of intimate experiences with natural areas and wild things—when he wrote: "People who care conserve; people who don't know don't care. What is the extinction of the condor to a child who has never known a wren?" A placed person will help us conserve our lake. The poet Gary Snyder believed that our relationship to place must be based on experience and information. He noted, in *The Practice of the Wild,* that many are alienated from their place and that they "don't even know that they don't 'know the plants'" of their environment. In an interview in *Poetry Flash* magazine, he also stated "So we have an

THIS PLACE IS WHERE A FAMILY COMES TO ENJOY THE LAKE.

environment whether we like it or not, wherever we are. And it's a relationship, place is a relationship like a marriage. Either you enter into that relationship, and it's very rewarding, or you deny that relationship, and you live in loneliness." A placed person will live a fuller life. Finally, Paul Shepard, ecologist and philosopher, went further. He believed that disconnecting from nature leads to madness. A placed person will be a more balanced person.

SENSE OF PLACE

The concept of sense of place is described in scientific journals on recreational use and environmental planning. In these writings sense of place is defined as a particular meaning attached to a specific place by a group of people. However, the concept is elusive. A place might be special to many residents and visitors, but to others the same place may not engender attachment or uniqueness. Sense of place requires the perception of belonging and of specialness about a place, whether biological, geological, aesthetic, or cultural. Aldo Leopold and William Whyte believed that the perception of beauty in a place is a requirement for developing a sense of place and for advancing an environmental ethic of place. In *A Sand County Almanac,* Leopold stated: "Our ability to perceive quality in nature begins, as in art, with the pretty. It expands through successive stages of the beautiful to values as yet uncaptured by language."

A sense of place encompasses what people know and feel about a place, as well as the activities and recreation that occur at a place. The landscape is an essential element of one's life. Place structures our thinking and activities. Place can become identity.

© Paul Radomski

Students and community members enjoy a warm summer day at the Memorial Union Terrace at the University of Wisconsin–Madison.

Sociologist Richard Stedman and psychologist Bradley Jorgensen surveyed lake home owners in northern Wisconsin to understand people's attachment to, dependence on, and identification with lakeshore living. They found that strong positive beliefs about lake home properties were common. They also demonstrated that the characteristics of the physical setting were very important in determining the meaning that people gave to a place and that degradation of a lake, its shoreland, and surrounding landscape resulted in a loss of sense of place and emotional displacement. People who liked a naturally vegetated lot and retained this vegetation were more likely to identify with, attach to, and depend on lakeshore living. Stedman found that seasonal residents were more attached to place than year-round residents and that seasonal resident attachment was enhanced by environmental quality and the escape of daily routine while year-round resident attachment to place was the result of the strength of a network of friends and view of the community.

Many researchers have observed that our culture tends to homogenize the human-made elements of place. Jane Jacobs wrote, in "Downtown Is for People," (printed in *The Exploding Metropolis*) that "a sense of place is built up, in the end, from many little things too, some so small people take them for granted, and yet the lack of them takes the flavor out of the city." Other design factors, too, can help contribute to a sense of place. William Whyte designed a simple scorecard to judge how well a city is doing to protect and promote a sense of place. The scorecard included the layout of the downtown, linkages, and the transportation configuration. He noted that vibrant cities with a sense of place had high foot traffic—the concentration of people was the critical factor. In *City: Rediscovering the Center*, he also stressed the need for clear edges: "The sense of place that a town can give is most felt by those who live within it. But it is also important for those who live beyond. A well-defined town with a tight core can give coherence to a whole countryside. It's a better place to live in if there is someplace to go to."

SENSE OF COMMUNITY

Knowledge, beauty, experience, and narrative can shape the sense of belonging and a sense of place, but rarely can these positives be separated from the notion of community. Even for those who go to the lake to escape the chaos of daily life, their neighbors, or their coworkers, there will be other people around. There will be a community, even if loosely structured or at a distance. And if that community is perceived negatively, the beauty of the place will not be appreciated, the paddling will not be so much fun, and the pike will be less interesting. Sharing the stories of the paddling and the pike and the trees will also be less appealing, and those stories are less likely to coalesce into place identity.

Community consists of the human complex (family, friends, neighbors, leaders) and the ecosystem (land, water, plants, animals). Including the natural world in the community can produce a sense of shared responsibility for the natural heritage, but this is not a simple act. The people within the community need to agree that frogs and trees are part of the community. However, using a highly inclusive and environmentally responsible definition of community can in fact rupture the community. Lawrence Buell, who studies the relationship between literature and the environment, noted in *Writing for an Endangered World*, "the more a site feels like a place, the more fervently it is so cherished, the greater the potential concern at its violation or even the possibility of violation." Buell wrote that we can be better stewards by recognizing that our place furthers our well-being, realizing that belonging matters more than possession, seeking a balance in our connectedness and sense of place with other views of place, and tolerating change knowing that all places are dynamic in time. In *A Sand County Almanac*, Aldo Leopold said, "When we see land as a community to which we belong, we may begin to use it with love and respect." If environmental policy is to make a real difference, people need to believe in a new definition of community, in a new land ethic. Whyte noted in "How the New Suburbia Socializes" that proximity is an important variable in determining friendship; "after all, you do have to come in contact with people before you can like them." Wendell Berry reminds us that to be members of a viable community we must understand how to be good neighbors. Berry stated, "a viable community is made up of neighbors who cherish and protect what they have in common." Around the lake, the lake is what people have in common, but they will have to be persuaded that it is indeed a shared asset and a shared responsibility; they will have to agree on the nature of the values, threats, and opportunities. In that sense, community building and community politics have to precede conservation. Curt Meine, conservation biologist and Aldo Leopold biographer, in *Correction Lines* said: "If a shared sense of place, value, and expectation is so elusive at home, how can we expect to find it at larger geographic scales, where the spectrum of needs and values is so much broader? Difficult as it is to be stewards of our own backyards, how shall we be so in our watersheds, our ecosystems, our nation, our earth?"

And what of the loon? Surely the loon is part of our lakeshore community. The loon watches boats pass by and reacts to our comings and goings, just as we do with our neighbors in town. Extending our community to grant membership to the loon also requires us to extend it to the water, plants, and animals on which the loon depends. Aldo Leopold was correct in stating, in *Companion to A Sand County Almanac:* "Conservation becomes possible only when man assumes the role of citizen in a community of which soils and waters, plants and animals are fellow members, each dependent on the others, and each entitled to his place in the sun."

From Olson's "The Laughing Loon" (*Listening Point*):

The canoe was drifting off the islands,
and the time had come for the calling,
that moment of magic in the north when all is quiet
and the water seemed hushed and waiting for the first lone call,
and when it came,
a single long-drawn mournful note,
the quiet was deeper than before.

GOVERNANCE

It is often noted that some of our most vibrant developed lakeshore places were created years ago without ordinances and with much lower regulatory complexity than exists today. That is not to say that there weren't constraints or principles at play. The economics and values at the time may have led to a more public-oriented outcome, or perhaps only the high-quality development manages to survive (i.e., survivorship bias). Today, we are often contradictory in our opinions of lakeshore development. We want to preserve the natural character of our lakes, yet we don't like limits to development. We do not want others to infringe on our freedom to enjoy our lakes, yet we want additional regulations on those who detract from our experiences. Our population continues to grow unabated, yet the miles of shoreline remain static; conflict is perhaps inevitable. Today's human demands exceed available natural resource supplies, whether for clear water, ducks, sunfish, or enduring lakeshore. The economics of supply and demand result in higher costs for lakeshore property, greater need for public boat launches, and more interest in private boat-mooring areas.

Our knowledge about the level of disturbance that our lakes can sustain without substantial consequences is incomplete. The consequences of a collection of small alterations on a system may be linear (raising at a constant rate with alterations), nonlinear (expressing tipping points), chaotic (displaying unpredictable behavior), or inconsequential. In addition, there might be time lags between the action and the consequence. Time lags make it difficult for the scientist to infer cause and effect and can lull policy makers into complacency on the need to address important environmental degradation problems that are ultimately just around the corner.

Our uncertainties about the effects of alteration vary by scale. At the lake level, we need to learn how much alteration will result in a loss of species, a reduction in productivity, or a catastrophic loss of resiliency. If we want answers at a regional, state, or provincial level, the uncertainties are even greater. The role of many state and provincial agency environmental permitting programs is to allow citizens to make some changes but to minimize the damage they cause. The activities these regulations allow are based on what is socially acceptable at the time. For example, many states regulate aquatic plant destruction. Minnesota's aquatic plant management permit program evolved from granting permits for water access to granting permits to improve recreation access, and their public waters works permit program has evolved so that some activities are no longer allowed or permitted.

Society makes most decisions based on the values of the day. At times, citizens will underestimate the consequences or the risks, and the resource suffers. Natural resource management professionals have a responsibility to alert citizens to the known and potential risks. At times, citizens will want the government to provide protections beyond what the facts might suggest. When confronted with these challenges, we need to rely on basic ecosystem management principles and facts to guide us to the answers.

The same applies to planning instruments: they evolve over time and reflect the culture and politics in which they originated. In some cases, governance and management of places appear constrained by past decisions, even when they seem illogical. Often in environmental and planning regulation, certain simple ideas and rules took hold early on, were replicated in many cases and places, and still mark the evolution of governance. Planning and environmental policy tools and their evolution can give us guidance on reinventing lakeshore living.

LOCAL AND REGIONAL PLANNING

We define local and regional community planning traditionally. Planning includes those processes and activities that governments use to control development, whereas many of the ideas presented in earlier chapters are examples of designing, a form of planning that is site-specific, design-oriented, and ecologically sensitive. We used terms like conservation design, green urbanism, or environmentally sensitive community design. We define community designing as essentially spatial governance, the coordination of policies and practices affecting the organization of space at the human scale. However, we cannot ignore existing traditions of planning, which are sometimes a hindrance for sustainability but also a set of procedures that can be navigated and used to leverage change.

Planning in this narrower sense, as in the activities that are usually labeled planning in North America, has a history that cannot be presented in full here, but it is useful to distinguish between land-use planning, comprehensive planning, and zoning. Land-use planning can mean virtually anything, as long as some kind of vision for the future land use in a community or region is involved. Land-use planning in Canada and the United States generally consists of an extensive public consultation process to form and document a long-term vision for how a community will develop land and protect important resources. Provinces and states have planning laws that set the ground rules for the planning process for cities, municipalities, counties, and regional planning agencies.

Comprehensive planning usually takes place at the local level, and it highlights the consistency and comprehensiveness of the vision: how are things related? If a train passes here, should we build a station, and if there is a station, does it offer the chance for a new neighborhood and maybe a park? However, comprehensive plans vary widely in content and effectiveness in obtaining planned outcomes. In some jurisdictions the plan consists of nothing more than goal statements on future development and natural resource protections. In others the plan includes analyses of development trends, assessment of transportation needs, recommendations on zoning districts, natural resource inventory maps, and the means to manage growth of development and achieve planned outcomes.

In many large urban centers, regional and metropolitan planning agencies exist to assist local governments to develop plans that provide coherence of land development across jurisdictions. Managing development growth for long-term financial prosperity of a community often requires policies and regulations that contain the physical space where expensive public infrastructure and services exist. Two common containment strategies employed by cities and regional planning agencies include greenbelts and urban growth boundaries. Greenbelts are lands bordering towns and cities where development is prohibited or substantially restricted. In ancient times these lands served as pasture. Today governments purchase or regulate greenbelt lands for recreation, farming, wildlife habitat, and environmental quality. Notable examples include Ottawa's Greenbelt, Boston's Emerald Necklace, and southern Ontario's Golden Horseshoe. In *The Last Landscape,* Whyte reviewed London's greenbelt and concluded that an effective greenbelt has to be a place that people can use and tightly defined using topography and water features. An urban growth boundary is a line that surrounds a city to designate where sewer and water infrastructure will not be extended and where high-density and farmland development is prohibited by zoning laws. Notable examples include Waterloo, Ontario; Portland, Oregon; and Toronto, Ontario. The use of this strategy is often controversial. Developers and landowners beyond the urban growth boundary often complain about the consequences to land and housing costs.

Zoning, then, is the practice of separating land uses at the local level and legally codifying this in a land-use ordinance and zoning map. The general idea is that zoning implementation must be consistent with the adopted comprehensive plan. In many communities, as Randall Arendt has noted, the application of zoning is not concordant with the comprehensive plan. In other cases, there is no comprehensive plan. These cases happen because many rural communities have neither a planner nor a comprehensive plan nor the expertise to ground such a plan in sustainability thinking and zoning ordinance. The generalized practice of comprehensive planning is a good thing for many reasons, but we cannot simply assume that it, along with zoning, translates into community design. We pay special attention to the zoning tradition now.

ZONING LAWS AND TRADITIONS

A zoning ordinance is an exercise of police power exerted in the public interest. The Constitution of Canada, under the Constitution Act, 1867, allows the provinces to make laws in relation to municipalities and property rights in the province. Article IV of the U.S. Constitution requires the federal government to guarantee to each state a republican form of government and ensures people will form a state's government and establish its laws. For many states, the laws that enable counties and municipalities to exercise shoreland use controls over local private development come from state statutes. These statutes give local governments both the authority and the responsibility to implement land-use controls that safeguard the community's future health, safety, and welfare and the local resources and values upon which they depend. Various courts have upheld the authority of state and local governments to regulate shoreland development through zoning ordinances to protect public resources. Because these ordinances play such a critical role in lakeshore health, care must be taken to ensure that they are well written and understandable, are uniform in application, and carry the support of clear public wishes.

In the United States and Canada, county and municipal governments generally administer shoreland-zoning ordinances; however, there are differences in this authority. In some places the state plays a large role in managing shoreland development. For example, the state of New Hampshire issues permits for construction (new home building, installation of sewage systems, etc.), excavation, and filling within shorelands and sets minimum standards for subdivisions and land use. State subdivision rules there include impervious surface limits, setbacks, stormwater management, and shoreline buffers. In other places, state and local government share responsibility in ensuring wise shoreland development. In Wisconsin, counties must have standards that meet or exceed the minimum state standards contained in state law, and they administer and enforce such shoreland ordinances. Similarly, in Minnesota the state government is required by statute to develop standards for the subdivision, use, and development of shorelands. These standards must be incorporated into local government (county, township, or city) ordinances, but local government controls can be more restrictive. (Minnesota is also one of many state jurisdictions to state that aesthetics alone is sufficient basis for a zoning control.) The shoreland zoning law of Maine requires municipalities to protect shoreland areas through adoption of shoreland zoning maps and ordinances. Maine provides municipalities with a set of minimum guidelines but encourages the local governments to adopt more protective provisions in their ordinances.

In some places the local government has primary or sole responsibility for managing shoreland development. The primary authority for managing planning and zoning in New York is entrusted to cities, towns, and villages, and these local governments rely on technical assistance from county and regional planning agencies, municipal organizations, and state government agencies. In Ontario and Quebec, planning and development decisions are made at the local level, while the provincial governments have regulatory authority for water quality and quantity. In places like Vermont, state agencies provide technical support to local governments on shoreland best management practices while requiring environmental review for large projects. In Michigan, the local governments formulate appropriate land-use regulations necessary for the protection of lakes.

Zoning ordinances vary in their complexity. The early zoning ordinances were simple documents with few details. The first zoning ordinance in the United States was developed in 1916 for New York City by Edward M. Bassett, the father of American zoning. This comprehensive zoning ordinance included height and setback controls, and it created residential districts that excluded other uses. This ordinance served as the model for the Standard State Zoning Enabling Act of 1926. Even though the Standard State Zoning Enabling Act was just a model ordinance and not an act of Congress, it became the foundation for local zoning ordinances across the county. The original zoning process elements from this model ordinance are still present in many local government ordinances today.

The evolution of zoning has led to increased zoning complexity. Today there are many types of zoning. The most common type, with its roots in New York City, is Euclidean zoning. This method of zoning separates land uses (residential, commercial, industrial, etc.) and provides dimensional standards for many of the uses allowed. The term "Euclidean" comes from the village of Euclid, Ohio, which was sued by a developer because the land he owned was reduced in value due to the village's zoning ordinance. In a landmark decision, Village of Euclid, Ohio v. Ambler Realty Co., 1926, the U.S. Supreme Court held that the village zoning ordinance was a reasonable extension of its police power to protect public health, safety, and general welfare. Another type of zoning, more recently adopted in the United States, is form-based zoning. Form-based zoning focuses on building form and the placement of buildings in reference to each other, to public spaces, and in the landscape. This zoning type, in particular the placement of buildings and streets in the landscape, is likely more appropriate for the types of development described within this book.

Within these two major types of zoning, ordinances often include performance-based criteria and incentives to give developers the flexibility to build projects that better target local government goals. New York City was the first to incorporate large incentives into their zoning by allowing more floor area in exchange for creation of additional public space. While performance-based criteria and incentives have benefits, the addition of these provisions into an ordinance increases administrative burdens and sometimes results in a lack of clarity in the approval process. Whyte saw the value of clear ordinances and believed that ordinances must be specific with regard to incentives and bonuses or "what you do not specify you do not get." Currently, shoreland ordinances often have common elements that include a process to address nonconforming development, criteria for shoreland overlay districts, lot size, minimum lake frontage, density limits, open-space requirements, placement and height of structures, placement and design of streets, shoreline buffers, land alteration, stormwater management, and impervious surface limits, subdivision provisions, and a procedure for granting and denying variances. The last in this list, variances, plays a major role in zoning.

Imagine the following scenario: You spent your hard-earned money on a lake lot where you want to build your dream retreat. However, the house you want to build is larger than that allowed by the local ordinance. A variance is necessary to build your lake home to your preferred plan. A variance is a process that governments use to give citizens the permission to break their own zoning ordinance rules for reasons of *exceptional circumstance.* Area variances are variances on dimensional standards (e.g., setbacks, frontage, height, etc.), whereas use variances permit uses other than those allowed in the land-use district (e.g., allow commercial use in a residential district). While variance laws vary from state to state, there are strong similarities as many of these state laws were based on the Standard State Zoning Enabling Act of 1926. Many state zoning laws still retain the basic concepts of the 1926 standard, including how variances are handled: "To authorize upon appeal in specific cases such variance from the terms of the ordinance as will not be contrary to the public interest, where, owing to special conditions, a literal enforcement of the provisions of the ordinance will result in unnecessary hardship, and so that the spirit of the ordinance shall be observed and substantial justice done."

Today, variances can only be granted when they are in harmony with the intent of the ordinance and the public interest. Granting of variances usually depends on determination of undue or unnecessary hardship, although "practical difficulties" is another term used in this context. Undue or unnecessary hardship, as defined by numerous state laws, generally requires three conditions. First, does the property owner propose to put the property to use in a reasonable manner? For example, you may not receive a variance to build your proposed lake home addition that does not meet lake setback requirements because the encroachment to the lake may be determined to be unreasonable due to other options being available. Second, undue hardship means that your predicament is due to circumstances unique to the property, not something you created. For example, if you built a lake cabin on the lot so that the place you wish to now build your garage or addition would be closer to the lake than the required setback, a variance might not be granted. However, if a small wetland was in the middle of your lot and you request a variance to build your cabin closer to the lake than the required setback, you might receive a variance because your predicament is due to the natural character of your lot. Third, if a variance is granted it should not alter the essential character of the locality. For example, you wish to build a large, tall home on the lake that would exceed the ordinance's maximum structure of thirty-five feet. If the character of development in the area includes mainly single-story homes less than thirty-five feet tall, it is possible that a variance would not be granted because a large, visually dominating structure might be perceived as altering the character of the area. Generally all three conditions must be considered and applied to each variance request, and the burden of establishing undue hardship rests with the person requesting the variance. Under most variance laws, economic or financial hardship alone does not constitute an undue hardship. In Minnesota, no variance can be granted that would allow any use that is prohibited in the zoning district in which the property is located. In other places, use variances are rarely issued due to the likelihood of changing the essential character of the locality.

A board of adjustment or board of appeals is often the authority for granting or denying variance requests. Board members must make difficult, impartial decisions that often are subjective. They must determine the facts, apply the criteria in the ordinance, examine alternatives, consider conditions, make a reasoned and objective decision, and document the process. The board can impose conditions to ensure compliance and to protect adjacent properties and the public interest. If a variance applicant fails to prove hardship (or practical difficulties), the board has no power to grant the requested variance. Despite this, at board of adjustment public hearings one can often hear property owners sharing an estimate of the compliance cost while arguing that providing a variance to the ordinance for a particular activity or development on their lakeshore lot will be inconsequential to the public resource. A granted variance is usually attached to the land, not the variance applicant. Board decisions have important consequences. Zoning ordinances and their compliance over time defines a community, and when done right they can increase the economic and natural resource value of an area.

Many government actions and regulations can affect property value. Some government actions provide value to property. Property value is determined by location, local amenities, and proximity to services and transportation infrastructure. Often government provides the infrastructure, and nature provides the most important amenities. Government might increase the local property values through creation of streets, sewer, water, and utilities or through effective protection of natural amenities and scenic quality. Riparian property owners benefit from zoning regulations that protect and preserve the valuable amenities associated with their lakeshore property. Regulations limit the disturbances to the natural shoreline and nearshore water that degrade public waters. Without shoreland ordinances, property owners would have to confront their neighbors to protect their private interests. Zoning controls are a much more effective mechanism than individual lawsuits.

Economic studies provide scientific support for the use of higher protective zoning standards for lakes. If zoning is not robust enough to protect the lake's environmental amenities (e.g., water quality, water clarity, scenic beauty, sense of place, and recreational opportunities), then existing and new development will diminish the value of the environmental amenity and the lakeshore properties. This is a common story in areas with insufficient shoreland development laws and rules. Too many people end up living along the lakeshore, and too many are unaware of their impact on the environment. The net effect is that many values are diminished, including property value and environmental integrity.

The cost to the local government of administering its ordinance is a public cost, borne by the taxpayers. These zoning costs are reasonable, as ordinances exist to protect the health, safety, and welfare of the community. The need for some regulation is universally recognized; the argument is over how much. Ordinances must be reasonable and applied with due process, and local governments cannot be arbitrary and capricious in their decisions and judgments.

The Fifth Amendment of the U.S. Constitution states, "nor shall private property be taken for public use, without just compensation" by the government. (Property is emphasized to reflect that it protects property possession and does not say private property value.) The U.S. Supreme Court reaffirmed the protection of private property in Armstrong v. United States, 1960, to keep government from "forcing some people alone to bear public burdens which, in all fairness and justice, should be borne by the public as a whole." Zoning authority is limited to what is written in ordinance and must be applied equally to all. All shoreland property owners must bear and comply with shoreland ordinances. Zoning exists to protect citizens from their neighbors, and society has determined that private property owners should not be allowed to use their property to advance their own self-interests without regard to the rights of others and the general public. The U.S. Supreme Court has also ruled that having to reduce the intensity of use of one's land does not constitute a "taking." There are different interpretations on this matter; however, the following statement from Justices Rehnquist and Stevens, in Penn Central Transp. Co. v. New York City, 1978, provides guidance: "A 'taking' may more readily be found when the interference with property can be characterized as a physical invasion by government than when interference arises from some public program adjusting the benefits and burdens of economic life to promote the common good."

A regulatory taking of property occurs when a government enacts regulations that deprive the owner of all use of a parcel or any part of a parcel (e.g., acquiring property via easement through regulation). A taking will also occur when a rule mandates open access or constitutes an invasion. Governments may adopt laws designed to protect water quality, including requiring shoreline buffers and other environmental protections, if they are in the interest of public health, safety, and welfare. In general, a regulation that diminishes property value alone does not constitute a taking. In Minnesota, however, a regulation that is designed to benefit a government enterprise such as an airport and results in a substantial diminution in value may be considered a taking. Lastly, in Koontz v. St. Johns River Water Management District, 2013, the Supreme Count ruled that government imposed conditions on land use must show a nexus and be in rough proportionality between the regulatory demand and the environmental effects of the proposed land use.

Legal scholar Joseph L. Sax investigated takings and land-use regulation issues. Sax identified the source of today's conflicts between property rights and public interests as the artifact of two conditions. First, there is often public resistance to long-term land development planning. Whereas planning and zoning staff might have good intentions for a community's comprehensive zoning map, elected officials frequently change the map to reflect the politics of the moment. The outcome is that land-use rules change, and this change disproportionately afflicts those property owners who have not yet developed their properties compared to those who have. Sax stated: "For better or worse, and in contrast to many other countries, America's reluctance to engage in long-range land use planning, and to diminish the autonomy of landowners, has produced a situation of late-stage public governance. That situation has, in some instances, thrust an undue portion of the burden of programs like open space, coastal, and biodiversity protection on the relatively few landowners who still have undeveloped or pristine land available. In the interest of fairness, such burdens should have been widely spread among the landowners in the affected community." The second condition that produces conflict is that zoning laws change. Zoning officials learn about the consequences of past development on public interests, and new information on how to protect public interests is continually available. For example, we now realize the importance of on-site stormwater management, and new developments may not be built as cheaply as those nearby due to updated regulations on stormwater management. Sax noted, "the endangered species situation provides the ultimate example; we wait until the species is about to become extinct, with the predictable result that the only regulated party will be the rare owner who has *not* destroyed the habitat value of his land, while everyone else has made their land useless for that purpose through use." Again, the cost of protecting a resource is unevenly spread. Together these two conditions have a tendency to burden those that develop later, first with unequal development rules, and second with unequal obligation in conservation. People react to these apparent inequalities.

Sax proposed various methods to address fairness with zoning systems. But first, he articulated a legal argument that recast the regulatory takings issue in a new light. Sax suggested that property owners should judge fairness in the context of both time and space rather than just in space. In other words, comparing your rights to those of others in the community with similar property (past and present), without regard to time, is too narrow-minded. Sax's review of U.S. Supreme Court cases suggested that recent decisions have relied too much on the context of space for determining fairness. Sax stated: "An acre proposed for development in 1954 is not the same as an adjoining acre slated for development in 2004. . . . The view that we take the world as we find it strikes me as much more our common experience than is the idea of trying to work out some equal allocation of use opportunities that stays stable over time. Indeed, our notion of the very identity of natural things (what they are and how they are to be used) has been distinctly time-contextual." He cited U.S. Supreme Court Justice Sutherland who wrote in the landmark zoning decision, Village of Euclid, Ohio v. Ambler Realty Co., 1926: "Until recent years, urban life was comparatively simple; but with the great increase and concentration of population, problems have developed, and constantly are developing, which require, and will continue to require, additional restriction in respect of the use and occupation of private lands in urban communities . . . the scope of their application must expand or contract to meet the new and different conditions which are constantly coming within the field of their operation. In a changing world it is impossible that it should be otherwise." He promoted the use of variances to address unfairness even though he stated, "in general, variances have been found to be too generously granted by local governments."

PUBLIC TRUST

Earlier, in a highly influential legal review that was published at a critical time for the modern environmental movement, Joseph Sax assessed the historical context of the contemporary legal principles of safeguarding public waters and aquatic resources. He used the term "public trust doctrine" to describe the legal framework that obligates state governments be held responsible to maintain public waters for the public good. The historical context for the public trust doctrine dates back to Roman and English law. The Romans believed that no one owned the rivers, sea, and shore—that all could use and enjoy these places. The English believed that the king was the trustee of communal property, such as the shoreline. These two views were merged in the United States, and the states were given public trustee status. In Canada, due to unique historical reasons and the role of the Crown, the public trust doctrine is not used. Given the doctrine's origins, its scope and force rests primarily in the protection of navigable waters, submerged lands, wetlands, and public access to public waters and aquatic resources. Today, many states have developed statutory laws specific to the protection of public waters. These state laws collectively constitute the water law of the state. State governments regulate shoreland activities that affect public waters and natural resources for their citizens, and their natural resource management agencies have permitting programs and rules designed to ensure long-term public good accrues from lakes. Each state has its own unique laws, and the scope of their public trust doctrines varies. Yet water law constrains the use of the water by citizens without any expectation of compensation. The public trust doctrine in the United States represents a reinvention of a timeless ethic for the treatment of communal resources, such as our lakes.

The vision developed in earlier chapters on sustainable design for rich lakeshore living, one that produces places that are pleasant to live in while maintaining the integrity of ecosystems, is possible within the legal and political context. Developing and implementing plans that are farsighted and comprehensive is a necessity. Rethinking and revising the concepts of public trust, justice, and fairness to include more ecological considerations and community goals will make the vision of ethical lakeshore living even more achievable.

CHAPTER 10

System Changing

LEOPOLD, OLSON, AND WHYTE WROTE ABOUT THE CAUSES OF MANY OF OUR LAKE-shore problems and strove to provide solutions. Leopold concluded that the protection of our native landscapes required an expansion of our ethics and that government could and should only play a limited role. Olson worked to engage governments to protect wilderness and wrote about the landscape in a way that inspired people to reconnect with natural parts of our world. Whyte worked with city, county, state, and federal governments to preserve and create public spaces for people and the rest of nature. Overall, each came to understand the need to change systems. There is no simple solution that will guarantee healthy lakes and vibrant lakeshore communities. Purposeful change of systems is very hard work; we must adjust our life goals and operating paradigms, and we must work together. We live in a time of exponential population increase, near our natural resource limits and carrying capacity, so we will need to move fast. Today there is less room for error and higher rates of conflict. Our problems are nested within other problems. Systems thinking can help us design and implement better solutions.

Systems thinking is an approach for estimating or inferring how actions or elements influence conditions of a system. As a process to solve problems, it also can be defined as viewing problems as parts of a system. Rather than reacting to a problem in a system with an action that changes nothing, or contributes further to the undesired problem, systems thinking provides a systematic approach to seeking a viable solution. It promotes an interdisciplinary approach in order to avoid a narrowly focused solution that may produce undesired outcomes or unintended consequences elsewhere in the system. Donella Meadows, in *Thinking in Systems,* provided a summary of systems analysis and suggestions on how to influence complex systems to obtain desired outcomes. One of the first steps in systems thinking is understanding how the system works. Only when one understands how a system works can one think about trying to change it. Meadows spoke of leverage points for change, defined as critical points to intervene in a system. She identified numerous leverage points and ranked them on their ability to transform a system. She revealed that finding a leverage point is difficult and, once found, changing it in the right direction is often counterintuitive. Leverage points include numbers, feedback loops, information flows, self-organization, rules, goals, and paradigms.

Numbers include zoning dimensional standards, water-quality standards, property tax rates, and other such parameters. Meadows noted that people tend to focus their attention on these parameters, but that they rarely have a lot of leverage. Critical numbers are not common in systems, as often a range of numbers will produce similar outcomes. For example, if 10 percent of the shoreline is in conservation easements, the benefits may be similar to those produced by 15 percent in easements. Often large number changes are necessary to produce big changes in a system.

Feedback loops are everywhere in ecosystems. Positive feedbacks include population growth; a high number of adult mosquitoes now means a high number of adult mosquitoes later (as the time from egg to adult is rapid). Negative feedbacks include predation; a high number of mosquitoes will likely lead to an increase in dragonflies and other mosquito predators and a resulting decrease in the mosquito population. Human-made systems include such negative feedbacks as supply and demand, price signals, and other self-correcting market forces. Our systems also include mechanisms that intentionally weaken positive feedback loops; these mechanisms include income taxes, inheritance taxes, and antitrust laws that attempt to weaken the rich-get-richer feedback loop. Feedbacks can be powerful leverage points, whether through strengthening of self-correcting feedbacks or moderating of positive feedbacks. In addition, adding information flow about a system adds a potential feedback loop. Self-organization is born when elements of a system change themselves. Culture and social norms change and evolve, and this has great leverage.

Rules and goals can have very high leverage when they are well defined, pragmatic, and consistent with current values. From the Constitution to zoning ordinances, rules can leverage major change. A new goal can change a system completely. The U.S. Clean Water Act, which set a goal of making lakes and rivers swimmable and fishable, changed how governments regulated pollution discharges and set water-quality standards. This new goal resulted in tremendous water-quality improvements across the country.

Finally, paradigms are the set of beliefs about a system, and they have the highest leverage potential within a system. Paradigms are the foundation of human-made systems, and they define our view of ecosystems. Paradigms are shared ideas of a culture, and they can be right or wrong. From the view that the world is flat to the view that our actions cannot alter the Earth's atmosphere, paradigms strongly persist until confronted repeatedly with their anomalies and with scientific evidence of a new paradigm. Paradigms are hard to change, but if one can successfully change a paradigm then the system will be reengineered, reconfigured, or reinvented.

We propose seven systems to change. In each, we identify a potential leverage point. We want to help people make beautiful places to live. Like ripples on the lake obscuring the rough and rocky bottoms, we may not see solutions to our challenges; however when the wind calms, details emerge. If progress is made in each of these seven systems, then better lakeshore places, friendlier neighborhoods, and more sustainable communities are truly possible. Working on the seven systems is working on the issues noted in earlier chapters, informed by the knowledge presented there, and inspired by the writings of Leopold, Olson, and Whyte. Even if much can be done with environmentally sensitive lakeshore property and community design, the analysis of the seven systems also shows that long-term sustainability of lakeshore living will also require interventions in other places and other policy domains.

1. REENGINEERING OUR SURFACE WATER MANAGEMENT

Failure to manage rainwater on individual lots and in neighborhoods is one of the primary problems negatively affecting lake quality. Proper development can reduce the negative consequences of rainwater while increasing property values. There are two ways to manage rainwater. The traditional way has been to move water off the land quickly. This approach uses stormwater sewers, pipes, and ponds. Unfortunately, only after this expensive approach was used across many areas did civil engineers find it did not work well. Often, the only outcome was the creation of larger problems downstream. This system does not scale well, and the treatment of water as a waste product instead of a resource discounts the importance of infiltration into groundwater and the value of the predevelopment water cycle. In addition to altering stream hydrology, an extensive underground piping network can potentially influence subsurface hydrology. In areas with substantial piping at or below the groundwater table, large amounts of groundwater may be drained away as it enters pipes at joints and cracks. The traditional way to manage stormwater is now seen as a failed system. Better management is to get the water into the ground near where it falls. This approach, often called low-impact development (LID), uses infiltration basins, rain gardens, grass overflow parking areas, grass swales, permeable pavement systems, parking lot infiltration islands, and overall less imperviousness. It also includes protecting natural areas important for water transport and filtering, such as wetlands, streams, and vegetated buffers near water. Utilizing this approach, runoff should match predevelopment conditions. In areas with ample precipitation, this approach reduces pollutants and nutrients entering into lakes. In areas with erratic precipitation, captured rainwater from impervious areas can be used for greening public places. This better way is small-scale and decentralized and treats rainwater as a resource instead of a waste product.

A simple performance goal for runoff reduction can be determined by retaining a runoff volume on-site equal to a certain amount of runoff volume from impervious surfaces on the site. A performance goal of one-half-inch runoff reduction per day may reduce some overflows and keep trash out of the water, but it likely cannot prevent an increase in runoff pollution. Runoff reduction over one inch per day might decrease runoff pollution and help maintain water quality but may not protect stream banks. A runoff reduction over two inches per day is needed to help maintain stream geomorphology and water quality, and runoff reduction of three inches or more per day may be needed to protect sensitive areas.

Homeowners can use rain gardens and other techniques to manage rainwater on their property. Rain gardens are landscaped areas planted with native wildflowers and other vegetation that soak up rainwater coming off the roof and driveway. The rain garden fills with water after a rain, and the water slowly infiltrates into the ground rather than contributing to the runoff problem. Cumulatively, numerous rain gardens in a neighborhood can have substantial positive environmental benefits. In addition, simply separating downspouts from impervious surfaces, reducing impervious surface area, and using permeable pavement systems can reduce runoff reaching streets and waterways.

Leverage Point

The leverage point for stormwater management is the introduction of a new paradigm by civil engineers and natural resource managers. These professionals are developing performance-based standards for treatment and infiltration of rainwater and technical manuals on the new methods. We can encourage local governments to move expeditiously to the new paradigm, and we can look at our own land to see where and how we can slow the water down and get it into the ground near where it falls. Trained professionals are available to assist homeowners with designing and implementing these rainwater management techniques.

The paradigm of getting the water into the ground near where it falls is not a new innovation. The famous Inca site of Machu Picchu, located in Peru, was designed in part with this paradigm. Machu Picchu is perched on a mountain-ridge site of twenty-five acres in a location of the Andes that received almost eighty inches of rain a year, and it would have eroded into the valley without critical thinking and engineering on rainwater management. This engineering work is still evident today. Before the estate and the town's buildings were constructed, the site was engineered to collect and hold rainwater, slopes were terraced, drinking-water fountains were installed, and a separate drainage system for waste was built. The terraced areas included sandy surface soils for infiltration over gravel and granite rubble from waste construction rock for subsurface rainwater storage. Machu Picchu has better rainwater management than most lakeshore areas do today. Our farm fields lose soil to wind erosion and water runoff, our tiled agricultural lands drain into rivers and lakes, and our urban centers pipe stormwater directly into lakes and rivers. We can do better, and the Incas are proof.

2. RETOOLING OUR WATER USE AND WASTEWATER MANAGEMENT

Developed counties have made great progress in wastewater collection and treatment. Rivers, lakes, and seas near large cities used to reek and kill with the overload of humanity's waste. Cholera killed millions of people in the 1800s. In 1854, Dr. John Snow, a London physician, determined that cholera was related to human waste contamination of drinking water. This information led to the construction and use of underground sewer systems to drain waste away from drinking water supplies before discharging it downstream of the city. By the late 1800s piped water and toilets became more common, with cesspools in backyards and sewer piped systems in dense cities. However, effective treatment of human waste did not occur until recently. The Clean Water Act of 1972 revolutionized wastewater treatment in the United States. In general, raw sewage is no longer discharged into public waters; however there are two big exceptions. First, hundreds of U.S. cities still have combined stormwater and sewage systems, so when a big storm sends water down stormwater drains, emergency overflow checkpoints send both untreated sewage and stormwater directly into public waters. Second, leaking sewers and illicit connections and discharges of untreated commercial and residential sewer wastewater flow into stormwater pipes or directly into streams and rivers.

There are two commonly used approaches to treat wastewater: subsurface sewage treatment systems and large centralized gravity-based systems with primary, secondary, and tertiary treatment technology. Both approaches have their problems. The problems with subsurface sewage treatment systems were discussed earlier. The centralized systems discharge treated water downstream and into rivers, lakes, reservoirs, and oceans. The effluent often contains high amounts of inorganic ions (e.g., chlorides) as well as new organic chemicals. The new organic chemicals come from industrial discharges, detergents, personal care products, pharmaceutical wastes, and unmetabolized, excreted compounds (e.g., caffeine or prescription and illicit drugs). For example, triclosan, an antibacterial agent currently used in hand soaps, and its poisonous chemical derivatives, including dioxins, have been found in lake sediments, and these organic chemicals can influence lake plants and animals. Other organic chemical compounds mimic natural hormones, and their introduction into downstream ecosystems can cause major problems. They may act as endocrine disruptors, mixing up the chemical messages that control development, reproductive changes, and other hormone-triggered processes in the body. The presence of these new organic compounds in aquatic ecosystems leads to a range of consequences, including the feminization of male fish downstream of wastewater treatment plants (now widespread and common from fish in the Mississippi River in

Minnesota to the Potomac River in Virginia), alteration of fish behavior, and the death of aquatic organisms.

Leverage Point

The leverage point for our subsurface treatment and large wastewater treatment systems is a change to the rules to require advanced treatment of effluent and to prohibit discharge to surface waters. First, a new generation of home or small-scale sewer systems are now available that provide advanced treatment of sewer tank effluent. These systems use recirculating sand filters, aerobic treatment systems, and organic/compost filters. For shoreland developments, it seems reasonable that this technology be utilized to better protect lake water quality. Second, prohibiting surface-water discharge from large wastewater treatment systems would trigger massive civil engineering innovations in water use and reclaimed water use within communities. This rule change would focus efforts on reducing unnecessary water use in homes and industry and would solve several long-standing issues on water sustainability and downstream pollution. This change may also focus efforts to advance biogas-to-energy projects. Few wastewater treatment facilities use anaerobic digesters to produce methane gas for generating electricity. Research has estimated that there is a large potential for biogas-to-energy projects at existing facilities. To contain costs and minimize future maintenance expenses, this rule would also require long-term designs to promote inward growth of cities rather than outward growth. There are several safe and important uses of reclaimed water. It can be pumped to infiltration areas for purification by natural processes and recharge of groundwater aquifers. It can also be distributed to landscape or agricultural irrigation systems, including conventional farm field systems and advanced urban hydroponic food systems. Expanding small hydroponic food systems and coupling them with wastewater systems could be an efficient method to grow substantial amounts of food within a city; the food-producing plants could use phosphorus from the wastewater, reducing nutrient pollution, and the water that was previously viewed as waste would now be an important resource for human and economic benefit.

3. RE-CREATING OUR GOVERNANCE SYSTEMS

Governments are increasingly complex. Objectives within different departments or agencies are often inconsistent, and goals are often separate or even competing. For issues related to lakeshore living or shoreland development, this can result in no clear purpose. Zoning laws are producing shoreland developments that the local communities often do not want. Governments have not stopped the decline of lake quality, and the government process rarely assists in the development of cherished places. Governments exist to provide security of food and rights through laws, as well as to allocate resources and to grant power to individuals and organizations. From banking regulation to zoning ordinances, a majority of the governed often prefers something between laissez-faire and the heavy hand of government. Governance is always imperfect.

A common suggestion to help state and federal water-planning efforts solve poor water management is to do watershed-based stormwater permitting instead of political boundary–based permitting. Such suggestions have been promoted since the 1800s, when geologist John Wesley Powell sought to integrate natural resources, communities, and institutions at the hydrographic or drainage basin scale. Powell's explorations took him to Wisconsin, Minnesota, the Mississippi

River valley, and the American West, including the first expedition through the Grand Canyon. Perhaps traversing diverse landscapes, along with his formal training in natural sciences, led him to advocate a concept that is still not fully addressed today in governance of natural resource use and exploitation. In 1890 Powell wrote: "There is a body of interdependent and unified interests and values, all collected in [a] hydrographic basin, and all segregated by well-defined boundary lines from the rest of the world. The people in such a district have common interests, common rights, and common duties, and must necessarily work together for common purposes." Powell concluded that people should be allowed to "make their own laws for the division of waters, for the protection and use of the forests, for the protection of the pasturage on the hills, and for the uses of the powers [created by the flow of water]."

Powell saw the critical need to coordinate the land and water management within the basin. While many states and provinces have watershed districts and numerous ongoing watershed efforts, a comprehensive state-regional-local institutional structure has not been developed or implemented. The lack of clear authority, planning, and policy at the watershed scale likely hinders this approach. In addition, the added complexity of yet another government regulatory layer may present untenable conditions. Where watershed planning has worked well it was due to large commitments by public and private organizations on small watersheds, clear goals, science that reassured participants that the goals were attainable and the implementation efforts were pragmatic, sufficient funding, and a set of champions that practiced an adaptive management approach.

Increasing energy costs and market forces will force us all to adapt. Close physical proximity to work, food, friends, and fun will gain importance. Whyte noted in *The Last Landscape* that government transportation and housing programs have great leverage in determining the quality of urban lakeshore living and "by subsidizing new freeways and peripheral beltways we can make it easier for people to move about within the outer area, but vigorous centers are not the less vital for this but the more, and a policy for dispersing their functions will fail." Economic changes will likely push cities to revise ordinances to promote compact, mixed-use, transit-oriented development and more efficient land-use patterns. It will be necessary to reduce government complexity and regulatory constraints to achieve these goals.

Today many governments fail to design their communities in any significant way beyond the scale of a project. The result is ad hoc neighborhood development, with outcomes reflecting the developer's short-term interests rather than the community's long-term needs. This is an underappreciated problem.

Leverage Point

Many natural resource agencies and local governments have responsibility for water quality, habitat protection, and economic development and planning. Regulatory complexity not only makes it difficult to get things done, it also offers opportunities for selective law enforcement and reinforces the negative images of planning and government in general. Rule complexity follows the law of diminishing returns. Regulatory complexity reduces efficiencies as both the regulated and the regulator spend time thinking about loopholes. There are few benefits of complex rules, and often regulators falsely believe that added complexity will reduce future problems when in fact those complexities create continuous administrative demands. The leverage point is to change the rules to reduce complexity and to be vigilant against degeneration to a complex state.

There are many benefits of fewer, simpler rules. Fewer rules create clarity, leading to political, economic, and legal advantages. Simple, robust rules are easier to implement than clever, difficult-to-understand rules. Simple rules are hard to evade. Fewer rules do not mean less powerful rules or less lake protection. Few, simple rules can still regulate complicated or complex systems. However, creating and maintaining simple, effective rules is hard. The common regulatory model is devolution of existing regulatory frameworks by adding exemptions and complexity. Rule exemptions are often the result of ad hoc responses to perceived issues of fairness or reasonableness. People want exemptions to avoid being regulated or for a competitive advantage. Regulators need to determine who really benefits from rule complexity. Rules can lead to more rules, so the regulators need to resist bloat and be vigilant to avoid the creation of fragile regulatory systems. For ordinances, structural regulatory reform has largely focused on the use of different types of zoning, rather than on confronting and dealing with the consequences of a complex regulatory framework. Zoning administrators are often distracted by their complex ordinances from the more serious effort of facilitating the development of beautiful places to live. Simple rules for the things we care about can be designed with systems thinking.

Governments could do several things to reduce the complexity of their zoning laws and processes. First, the state, provincial, and federal government regulations could be simplified by focusing on only essential and highly effective natural resource protection provisions while allowing some flexibility to local governments. This will require a thoughtful reset of regulations. Second, local governments could reduce the complexity of their land-use tables and other provisions of their zoning ordinance to ease administration. If local governments restructured their complex land-tables, Donald Elliot, in *A Better Way to Zone,* stated that they may address several issues: (1) improved consistency and predictability in judgments by providing more flexibility in uses and fewer districts; (2) increased mixed-use and transit-oriented development by using just three district types (pure residential, mixed-use, and special purpose); (3) greater flexibility for citizens and developers in redevelopment by specifying ranges of allowable outcomes.

Third, governments could simplify by doing lakeshed designing up front. This situation is ideal to implement the principles outlined in previous chapters: a vision for the lake and its development can most profitably emerge independent of any specific proposal. If that proves impossible, one can imagine a strategy where local governments wait for initiatives by private actors and then develop a comprehensive plan for both the development and the lake; communal review would determine whether the plans followed the criteria for sustainable and just design. *In other words, a few rules and one planning-as-design process can replace the application of many rules in absence of a design.* Approval of the design would imply compliance with a series of ordinances and laws. In places where this approach is politically not feasible, local governments could make other changes to reduce the complexity of their planning and zoning processes and orient them more toward sustainability. For example, regulations could be modified to allow local ordinance flexibility. Jeff Schoenbauer, a Minnesota landscape architect, suggested that local governments take a leadership role in development by working more collaboratively with developers. Specifically, a local government works to achieve sustainable development by leveraging its traditional regulatory authority track, which usually accepts lesser outcomes, with the use of a collaborative track to get collective results (a new and more environmentally conscious variation on the traditional planned-unit development). This approach allows greater creativity and flexibility but requires a vision for a development or redevelopment site in the context of a thoughtful comprehensive plan, adequate staff capacity, and political and developer will. Given that most large developments are already negotiated, that issues on large sites are difficult to anticipate, and that innovations on large sites are often critical for a city or community, this approach is reasonable and has merit. Local governments would need to negotiate to obtain their community's design and to use the principles outlined in Part 3 (i.e., asset-based design, asset creation, and ensuring connectivity). The approach also requires that the negotiated outcome meet or exceed the intent of the appropriate development standards (the regulatory track). Several municipalities have already used a collaborative track; for example, the city of Lino Lakes, Minnesota, used the private development process along with public acquisitions to begin creating an extensive park and trail system.

There are other places for structural regulatory reform. The best rules are strong because they are simple. A simple rule based on a good goal, principle, or paradigm is easier to implement than detailed prescriptive rules. If people know the why and it is consistent with their values, then it is often easier for them to accept and comply. For example, it was estimated that in the United States lawn covers more than three times as much land than does any irrigated crop, and lawn watering accounts for up to 75 percent of the total residential use of valuable drinking water. So a simple rule for a city water-conservation ordinance might state that residential lawns can only be irrigated with household gray water (conservation goal: promote the reuse of water and reduce the use of city water, treated at taxpayer expense, for nonpotable purposes). Related to #1 above, a simple rule for city ordinances might require that residential roof runoff be diverted away from connected impervious surfaces and directed to infiltration areas (water-quality conservation paradigm: get water into the ground near where it falls).

Finally and most importantly, governments will need to radically adjust their ordinances to let go of many existing rules while redirecting and focusing efforts toward the creation of public places along lakes and streams and toward street layouts that developers must follow (see #6 below). This change will require designs that deliver outcomes, not more unimplemented ideas. When we do not design we are gambling with low odds at retaining and creating quality places. We pay higher taxes for poor land-use decisions that support inefficient infrastructure, and we live with degraded environments. Today, neighborhood development occurs with little planning, a heavy dose of ordinance and code, and a developer with good short-term vision. A better alternative is a community with an unambiguous public-oriented lakeshore and lakeshed plan that sticks, simple rules to protect only those essential resources, and developers given extensive freedom within the boundaries of the plan.

4. RETHINKING INVASIVE SPECIES MANAGEMENT

A considerable amount of energy and resources is now dedicated to invasive species management in our lakes. However, the benefits of this effort have not been thoroughly questioned. To re-create our governance system to be more effective in implementing lakeshed planning, we will have to ask our governments to rethink invasive species management. Resources now dedicated for mitigation of short-term negative consequences of species migration could be redirected to efforts ensuring effective long-term lakeshed management practices on public and private properties for the protection of water resources.

Invasive species issues have become socially laden, and even the language associated with invasive species management often includes strong connotations and militaristic metaphors. The term "invasive" means that an organism is not originally from the area or that its migration was human assisted. On a climate-changing and human-dominated planet what is a species's "natural" range? Species are neither good nor bad based on where they originated. Leopold noted in *A Sand County Almanac,* "In the end every region and every resource get their quota of uninvited ecological guests." Why do we label species bad if they are successful in a new area? Is it because they threaten something that we value? Or does an invasive species reduce ecological potential, such as productivity?

Biodiversity can be defined two ways: the number of species in a specific area (α-diversity) and the turnover of species across space or between habitat (β-diversity). Community diversity, as measured by β-diversity indices, refers to the variability of communities within a region. Fish transfers and introductions have played an important role in homogenizing fish communities across lakes and decreasing β-diversity. Fish management agencies played the largest role in this homogenization, but anglers, including Sigurd Olson, also contributed. In 1940, Olson obtained permits to introduce smallmouth bass into lakes of the canoe county of Quetico-Superior. Olson's introductions were successful, and it is unknown whether or not he had regrets about this invasive species introduction.

People may object to invasive species for purely aesthetic reasons. The similarity of plants and animals across regions or continents can diminish a sense of place. A diverse ecosystem, at the moment dominated by a nonnative species, may not hold the same appeal for its residents and visitors. Alternatively, invasive species may be enjoyed for these same reasons. Sigurd Olson transplanted yellow iris (*Iris pseudoacorus*), a European species, from the St. Louis River near Duluth, Minnesota, to a little wetland near Ely and to Listening Point on Burntside Lake. He wrote of his motivations and the benefit of adding this nonnative species to the area in "Fleur-de-lis" in *Listening Point;* much of his reasoning was romantic or sentimental. However, because yellow iris competes with native shoreline plants, many natural management agencies in North America have labeled yellow iris an invasive species and made it illegal to introduce it into lakes and rivers.

Ecosystem functions may change with the arrival of an invasive or alien species. Who is the judge of the merits of a changed ecosystem function? Are we biased in our judging because we are generally biased against change? As Mark Kurlansky, in *Cod: A Biography of the Fish that Changed the World,* stated: "Man wants to see nature and evolution as separate from human activities. . . . But man also belongs to the natural world. . . . If cod and haddock and other species cannot survive because man kills them, something more adaptable will take their place. Nature, the ultimate pragmatist, doggedly searches for something that works. But as the cockroach demonstrates, what works best in nature does not always appeal to us." We are also biased in the control of some invasive species. The most common invasive species present along our developed shorelines is Kentucky bluegrass (*Poa pratensis*). The presence of this species results in increased runoff reaching our lakes and leads to lower water quality and associated problems. Yet how does this species fit into your natural resource management agency's invasive species program?

Many invasive species management plans are based on the belief that our ecosystem interventions produce better results than those that would arise from species interactions over time. These interventions include the use of chemical treatments (herbicides, insecticides, and other pesticides), mechanical removal, and biological control. Sometimes our interventions result in unintended consequences. For example, pesticide treatment efficacy of curly-leaved pondweed (*Potamogeton crispus*) has been highly variable, and some treatments have resulted in diminished native plant communities and production of nuisance algal blooms. The remedy should not create worse conditions. Curly-leaved pondweed has been in some of our North America lakes for over one hundred years. Sometimes our interventions are incredibly successful. Biological control involves the introduction of a predator or disease vector. This method has risks but can be highly efficient and effective. Purple loosestrife (*Lythrum salicaria*) is an ornamental perennial that moved into wetlands, drastically changing the nutrient cycling and hydrologic characteristics of these ecosystems. After rigorous testing, four leaf-eating beetle species were approved for use in the United States, and they have effectively reduced purple loosestrife dominance in wetlands.

Every year, millions of shipping containers are moving from country to country and place to place. The consequences of this worldwide travel include the inadvertent movement of organisms. However, nonnative species are also moved intentionally. For example, five Asian carp species now occur in North America. In addition to the common carp (*Cyprinus carpio;* brought over as a food fish), the others include grass carp (*Ctenopharyngodon idella;* brought over in 1963 to control aquatic plants), bighead carp (*Hypophthalmichthys nobilis;* brought over in 1972 for aquaculture purposes), silver carp (*Hypophthalmichthys molitrix;* brought over in 1973 for aquaculture), and black carp (*Mylopharyngodon piceus;* brought over in the 1970s for aquaculture). All but the black carp have dispersed across the country. Governments will spend millions of dollars controlling carp after allowing or simply failing to regulate their importation.

To many it appears that the current management approach for exotic or invasive organisms is unsustainable or ineffective. Once an organism finds a place in an ecosystem and begins to successfully reproduce, it is often hard or impossible to remove or even control. Many nonnative species have integrated into our ecosystems, including our wild ecosystems. We advocate managing ecosystems with their native coevolved diversity when and where we can; however, we must take management actions that recognize that our ecosystems are often made up of a mix of native and nonnative species and that they will continue to be so. We also need to adapt to today's reality and manage the root causes of negative environmental changes, such as eutrophication and other ecological disturbances. Aldo Leopold was concerned in "What Is a Weed?" (printed in *For the Health of the Land*) about antiweed propaganda in the agriculture community and the failure to address the root causes of change, such as needless disturbance of soil and other misuses. Natural resource management agencies can and should promote behaviors that reduce human-assisted invasive species movement, and it may be possible to mitigate some of the consequences of an invasion. However, we need to learn to live with some of the organisms that we have intentionally and unintentionally moved around. It may be naïve to believe that each species has only one unchanging place to which it belongs.

Leverage Point

For the system of invasive species management, a leverage point to change might be the goals of the system. In addition to the goal to reduce human-assisted migration, three goals could be added. First, if an invasive species does not substantially threaten what we value and its role in the ecosystem is similar in function to those species already there (based on empirical evidence), then there appears to be little economic or ecological reason to control its population. Curly-leaved pondweed might only be managed in the context of individual nuisance water-access conditions, but sea lamprey (Petromyzon marinus) and Kentucky bluegrass likely would be controlled to reduce their populations and spatial extent (as these species dramatically change ecosystem functions). Management of a species should not be dependent on origin or order of colonization. Second, natural resource agencies could prioritize places where they wish to re-create and maintain the native coevolved diversity. These areas might include national parks and scientific and natural areas. This does not mean that invasive species are not present in these locations, but they are controlled to low abundance, and the ecosystem functions much as it did in the past. Habitats, sedimentation, nutrient cycling, and trophic level structure all look and function similar to how they did before invasive species were present. Lastly, making more rational choices on living with nonnative species is key.

The third goal includes reconciliation. Ecologist Michael Rosenzweig's concept of reconciliation ecology is an approach to find a balance between human needs and wildness while increasing diversity in our domesticated places and conserving species diversity in our wild places. With regard to species compositions, we need to care more about biodiversity than managing only for compositions from the recent past. A reconciliation goal is a more sustainable approach than many natural resource management agencies now practice, whose actions are often primarily driven by unrealistic goals of preservation or restoration to natural conditions. It does not mean that these agencies give up on reducing human-assisted migration, but it does mean that they might give up on the treatment or management of particular nonnative species. As Mark Davis, another ecologist, in *Invasion Biology* stated, "in instances where nonnative, and even non-native invasive, species are not causing a significant harm . . . altering one's perspective is certainly much less costly than any other sort of management program." Reconciliation recognizes that the war on invasives cannot be won, that the migration of organisms will continue, and that we can admire the beauty of all plants and animals regardless of how they arrived in our dynamic ecosystems. Stop the war and start working for nature and the conservation of natural features in our communities.

5. RECONFIGURING CAPITALISM AND EXPANDING MARKETS

There are two commonly recognized theories regarding natural resource management dilemmas. The first, developed and named by Garrett Hardin in 1968, is the "tragedy of the commons." Here, open access or common-pool resources, such as fish stocks, lakes, rivers, biodiversity, and aquifers, are overconsumed, and the costs for overusing the commons are distributed to the larger community (i.e., a negative externality). Elinor Ostrom won the 2009 Nobel Prize in Economics for her studies on how people have worked together to minimize occurrences of the tragedy of the commons. The second theory, developed and named by Christopher Lant and others, is the "tragedy of ecosystem services." Here the dilemma is the underproduction of ecosystem services. Ecosystems provide goods and services that support human welfare. Ecosystem services have been defined as "a wide range of conditions and processes through which natural ecosystems, and the species that are part of them, help sustain and fulfill human life," and examples include water purification, soil formation, nutrient cycling, and flood regulation. With the tragedy of the ecosystem services dilemma, the benefits of producing ecosystem services go to free riders (i.e., positive externality). For example, a landowner with a wetland that filters agricultural runoff has an ecosystem that exports clean water (a service) to downstream users who do not pay for that service.

Our existing economic systems generally fail to value ecosystem services. Economists track how money flows in the economy. Limnologists track how carbon, phosphorus, nitrogen, and other basic elements cycle in lake ecosystems. These natural systems become less resilient with greater demands. While money determines the fate of businesses and the purchasing power of individuals, the basic elements and the energy necessary for chemical reactions determine the fate of species and civilizations. The leading world economies have failed to effectively price public goods, such as clean air and water. Today these economic systems do not value what the painted turtle and loon value, and from their perspective our economies must appear to operate as pyramid or Ponzi schemes. At the moment, money is a weak proxy for the real world trading and cycling of the basic elements of life and our natural resources. It need not be.

It is recognized that our economic systems and markets work to the advantage of those that take less than full responsibility for the pollution they create, the exploitation of a public resource, or the damage their actions have on nearby natural resources. Most governments have failed to create markets for obvious environmental goods and services. One role of government is to set fair and reasonable rules for markets, and governments need to structure markets so that they function for the benefit of the community—not the polluter, the exploiter, and the destroyer of economic and environmental value.

Leverage Point

Managing for healthy watersheds and clean and adequate water sources will increasingly require understanding of ecosystems, hydrology, limnology, economics, and human needs. Successful watershed management to protect groundwater and surface water will likely require regulation of both land and water resources. However, given the limits of regulation, we will need to employ multiple tools, including market-based tools. These tools include incentives, certification, and private or public payment for ecosystem services. The premise of ecosystem services is that healthy natural systems provide a multitude of benefits to humans; some of these benefits have a marketable value and some do not, but all improve the human condition. There are many examples of payment for ecosystem services systems around the world. They include simple and pragmatic solutions like the private payment for ecosystem services in Tanzania where communities that regulate their land use to minimize soil erosion receive private payments from a downstream water supplier; this system ensures the continued delivery of the ecosystem services (i.e., clean water) from their lands. Complex and effective public payment for ecosystem services solutions include the one that New York City uses to ensure safe drinking water for ten million people by protection and restoration of the Catskill watershed. In order to address documented natural resource problems and an energy crisis, governments will need to be forward thinking and continually adapt. This may require counting or accounting for the value of ecosystem services in our decisions; strengthening the link between economics, research, and management; raising our awareness about the uses of payment for ecosystem services; and building our capacity to deal with this approach.

Market-based, private natural resource management may be an effective way to manage public goods that are currently treated as externalities. Governments may set fair and reasonable market rules so that ecosystem services are managed for the benefit of citizens. Private-public partnerships using nonprofit corporations and cooperatives with government contracts may promote the public interest, since these organizations are under the control of their membership and could be responsive to natural resource sustainability. State, provincial, and local governments may need policy and legal frameworks to experiment with this approach. The employment of nonprofit corporations and cooperatives in the management of public resources within watersheds, whereby these organizations charge property owners for use of ecosystem services and provide payment to those property owners that protect ecosystem services, may be an effective leverage point. This expansion of our capitalistic system adds critical negative feedback loops. Sample markets include markets for water retention for flood damage reduction, markets for point and nonpoint water discharge for water-quality protection, and markets for water use for maintenance of river flows and groundwater levels, which support base river flows and lake levels.

Reconfiguring and expanding our capitalistic economic system with price signals for ecosystem services is consistent with our values and traditions. It is important to pay as we go, and everybody should pay the full cost of his or her activities. We should have markets for ecosystem services. An expanded economy may help create a system that is fair and equitable, and it may provide an important feedback loop for lake protection.

6. RENOVATING OUR CITY SYSTEMS AROUND LAKES AND RIVERS WITH DESIGNS

The high cost of lakeshore shoreland means that it often gets subdivided into many small parcels for high-value lakeshore residential properties. However, the long-term best use, both economically and ecologically, is that many of these lakeshore areas be protected or converted back into undeveloped natural areas and parks. Every city, especially those with lakes and rivers, should have a connected network of parks, pedestrian parkways, and natural areas. This is how cities were designed and constructed in the past. Unfortunately, most of today's cities are no longer designed but develop without serious long-term thought about either grey infrastructure (sustainability of sewers and streets in outward growth areas) or green infrastructure (livability with parks and natural areas).

Modest cities become great cities with the addition of connected parks, parkways, and natural areas—witness New York City's Central Park, Minneapolis's Chain of Lakes, Chicago's Wilderness, and Boston's Emerald Necklace. The most valuable land in the world is that land surrounding Central Park; the most desired locations for urban living in Minneapolis are near the parkways and lakes. Chicago's Wilderness is a network of over 370,000 acres of protected lands and waters in a landscape that includes nine million people. Over 250 organizations are working together to use conservation tools (e.g., conservation easements, conservation subdivisions, the paradigm of getting the water into the ground near where it falls) and land acquisitions to connect Chicago's green infrastructure in a way that provides better outdoor recreation, flood control, water infiltration and purification, and lake and stream protection. Boston's Emerald Necklace constitutes half of the city's park space. The Emerald Necklace is a continuous seven-mile chain of parks connected by parkway and shoreline corridors and comprises over 1,100 acres.

Patrick Condon, in *Seven Rules for Sustainable Communities,* identified several important concepts that designer Frederick Law Olmsted used to make Boston's Emerald Necklace successful. First, and most importantly, lakeshore and streams became public rights of way; they are utilized as the front door for the community rather than the back door of private property. This approach makes the lakes and streams public amenities for all to enjoy. This concept requires placement of streets with undeveloped shorelines on one side and developed areas on the other side (i.e., single-loaded streets). Second, natural areas were used to contain and define neighborhoods. A park and connecting parkway system can create a sense of place by defining boundaries or edges of a neighborhood. Third, wide tree-lined boulevards contribute to the beauty of the area, as well as provide water infiltration areas and other ecosystem services. Fourth, Olmsted designed the parkway system and the associated street network to be infinitely expandable. This design concept is critical. Finally, the Emerald Necklace is continuous and connected. This allows people to get close to nature, either by walking or biking, and provides an alternative transportation system within the city. These successful examples can inspire cities to think big with regard to public places around lakes and rivers.

Leverage Point

To create a community system of connected parkways and lakeshore, the leverage point is to change the goals of lakeshore development. We need to do less zoning and more community designing. Recall that we see community designing as essentially spatial governance, the coordination of policies and practices affecting the organization of space. Local governments could facilitate environmentally sensitive, human-scaled community designing processes. Substantial amounts of lakeshore within cities could be public for all citizens to enjoy, and all citizens would be able to access and enjoy a parkway near their homes that connects to a larger system. Communities should require future development and redevelopment to conform to an expandable design. Regional designers, not developers, should determine street and natural area locations, as this task, given its long-term importance, should be the responsibility of the municipality or regional authority. This means laying out where future streets and roads will exist at approximate times and where existing streets will be relocated when rebuilding. The layout should produce a mix of meandering streets to ensure that lakes and streams become public rights of way and sensitive areas are avoided. In addition, streets and streetcar corridors should be created in a grid pattern in relatively flat upland areas to increase walkability and urban vitality. This change in responsibility will ensure conformity with a city's or region's comprehensive plan. We need to demand this action. Holly Whyte said in a 1986 interview: "If I'm lecturing to a civic or planning group, I say you guys aren't asking for enough. The developers will give it to you, but you have to stretch for it. It's for their own good. They might not know that, but in the long run, it is."

We need large-scale designs. Rather than dealing with the minute details of land use and residential and commercial property design, cities and regional authorities could redirect their efforts toward identifying sensitive natural areas for protection and lakeshore for public rights of way, regulating the design of public spaces and the placement of buildings in relation to public spaces, and declaring future street layouts at the city and regional scale. A systematic approach aimed at retaining important environmental benefits while reducing interference between competing land uses—such as the one presented in this book—will be critical in these efforts, and there are many conservation prioritization tools that can be used.

In places where governments suggest that developers retain public functions, such as street and pipe installation and maintenance, the citizens in the community should not subsidize the development. For too long developers have realized private gain at public expense. If developers wish to design a street layout inconsistent with the expandable design, they should first pay a development impact fee that reflects the true regional cost of each building on public services and resources (e.g., construction and long-term maintenance of streets, stormwater management, parks, water and sewer, fire, police, and ecosystem services). The development impact fee could be based on the marginal costs of growth, such that the fees would be greater for areas further away from urban centers.

These changes will require big thinking by landscape architects and designers and flexibility from local government administrators and natural resource management officials, but most importantly they will require promoters and champions for this valuable cause. In the past a single person pushing hard to ensure that a city placed lakes and rivers at the center of its residents' lives led to significant improvements in community design. Today one dedicated person could do the same to create a sense of place.

Horace Cleveland, the landscape architect who designed the interconnected series of greenways and parks centered on the lakes and the Mississippi River in Minneapolis stated in 1883, "Look forward for a century, to the time when the city has a population of a million, and think what will be their wants. . . . They [the wealthy elite] will have wealth enough to purchase all that money can buy, but all their wealth cannot purchase a lost opportunity, or restore natural features of grandeur and beauty, which would then possess priceless value." In 1880 the population of Minneapolis was about fifty thousand people. Over a century later, the population of the Twin Cities metropolitan area is over 3.5 million people, and the vision and dedication of Horace Cleveland in acquiring and protecting lakeshore is truly enjoyed by citizens of Minnesota. Cleveland's words are as true today as they were in 1883. It is our turn to take action!

7. REINVENTING OUR LAKESHORE LIVING

There is no one right way to live, and one should live how he or she thinks is best. Many lakeshore property owners understand the water-quality and habitat benefits of shoreline buffers, and they've restored their shoreline to natural conditions. They re-created their notions and ideals of a healthy shoreline. Likewise, many people have reset their views with respect to treatment of rainwater on the landscape and the importance of striving for sustainable living. Unfortunately, some social norms are clearly creating problems for our communities and our natural environment. In this context, it seems reasonable to call for civic responsibility and to question certain ethical systems. Aldo Leopold was concerned about how we ought to live on the land and provided advice on living that was derived from his passion for nature. In a manuscript from 1946 on conservation he stated: "The average citizen, especially the landowner, has an obligation to manage his land in the interest of the community, as well as his own interest. The fallacious doctrine that the government must subsidize all conservation not immediately profitable for the private landowner will ultimately bankrupt either the treasury, or the land, or both. The nation needs, and has a right to expect, the private landowner to use his land with foresight, skill, and regard for the future."

Ethics do not spring from the soil or bubble up from the lake depths, and Leopold noted in *A Sand County Almanac,* "We can be ethical only in relation to something we can see, feel, understand, love, or otherwise have faith in." Leopold developed the land ethic, and his essays in *A Sand County Almanac* elucidate how the philosophical foundations of the ethic came to be. Today, most people have incorporated the land ethic into their bundle of beliefs and morals. Most people believe in the lakeshore living principles of Aldo Leopold. But what about those who do not? How do we bring out the best in people to live compatibly with this ethic? And to what extent is this quest to impose ethics ethical?

Leopold believed that an ethical citizen would restore an ecosystem if he or she truly understood its wounds. However, the scaffolding around the land ethic has not yet been adequately built up, in part due to a lack of practice. We can begin by practicing how to restore native shoreline vegetation, and we can have our children and grandchildren practice a land and water ethic by playing with terrariums and aquariums. As we practice the land ethic, the new social norms will become self-evident to others in our community, leading to further and better practice. But still, the question remains whether the ethic can be imposed. We argue that, whatever belief system one adopts, it is reasonable to be confronted with the consequences of one's actions. And we believe it is reasonable for a community to impose its standards, both legal standards and ethical norms, on individuals who harm the public good.

Leverage Point

To make good things happen around our lakeshore communities, the critical leverage point is self-organization. When lakeshore homes and developments were fewer, smaller, and seasonal, there was little need for rainwater and wastewater management. But as development became more intensive and extensive, the need for innovations to address public health, safety, and welfare increased. Today citizens need to adopt new approaches on water management, natural area protection, and lakeshore development. Information and education help but are not enough. Clear and concise guidelines help but are not enough. Laws and ethics must be part of the solution. Only with the supporting ethics can we cultivate a sense of civic responsibility and an encouragement of community environmental activism.

© ANDREA LEE LAMBRECHT

WE ARE DRAWN TO THE SHORE, BUT HOW WE TREAT THE LAND DETERMINES THE QUALITY OF THE LAKE.

CHAPTER 11

Our Lake, Our Responsibility

ECOLOGICALLY, EVERY PLACE IS UNIQUE, AND HUMAN HISTORY AND CULTURE ADD layers to that uniqueness. Water is vital ecologically and culturally. Today, ecosystems and ecological values are under increasing threat, and human nature is such that new threats continue to emerge. Living close to the water's edge disturbs one of the ecosystem's most vulnerable spots, but it is also central to our self-understanding. Reinventing our lakeshore living requires rethinking place identity and shifting cultural identity in the areas we inhabit. Leopold, Olson, and Whyte were keenly aware of this, and their writings, each in a different way, shed light on the complexity of such endeavor, on the complexity of the relations between individual, community, and place. They suggested, sometimes plainly, sometimes subtly, how we might resolve the issues we've created, issues that strongly affect the lakes we love.

Leopold was primarily involved in management of areas with a history of tragic desecration and abuse, with a variety of stakeholders pushing in different directions. He advanced a land ethic in the hope that he would see immediate improvement in our treatment of nature. Olson referred often to "wilderness" and was active in its protection. He was very inclusive in his interpretation of wilderness, believing that a wilderness experience can take place in the backyard or in a city park. Whyte personally experienced the legacy of suburbanization and its implications for nature. He promoted cluster developments and conservation easements, and he used science to improve city public spaces. Their combined experience makes it unequivocally clear that change is possible, that shards of nature and wildness can have great value, and that careful management can promote ecosystem diversity and resilience.

Devising community design strategies that support the insights of our writers requires a cautious approach to the landscape; much of their writing is an argument for context-sensitive design, for carefully embedded interventions. It is also an argument for uniqueness. Understanding the unique character of a place can lead the way to a design that might actually mean something for the people living there, inspiring them to protect the quality of space, life, and ecology. Continuing our current habits of parceling up the lakeshore, of creating individual enclaves on the water, will not work. Building on actual natural and cultural assets is crucial in the creation of place identity. Maintaining place identity can only occur in a community, so the design must envision and encourage community.

Leopold, Olson, and Whyte wrote often of combining uses in appropriate settings and separating uses in others. The same applies to the coexistence and separation of humans and the rest of nature; a deeper knowledge and appreciation of the local landscape can help to devise strategies to preserve wildness here, restore it there, or in some places aim for a more cultural landscape and plan. Landscapes help form places. Their writings also argue for a

143

comprehensive approach: working toward place identity on the smallest scale can only work if there is a concerted effort to design at larger scales. There must be not only a design for the home, but a vision for the whole lake, the city, the region. Our writers also confront us with the uneasy fact that things become more appreciated when they are rare. The environmental movement gained momentum when much was lost. But people also get attached to what they appreciate and understand. Giving people the opportunity to know a place better, to experience a place in a different way, to enrich that experience and multiply the perspectives on a place, not only creates place identity but can also deepen the attachment to that place. Olson explained in *Open Horizons,* "only if there is understanding can there be reverence, and only when there is deep emotional feeling is anyone willing to battle."

Showing people some of the characteristics that make a place unique, like its vulnerable wetlands or rare wetland birds, can help foster their appreciation, attachment, and stewardship to that place. Bringing people too close endangers the qualities observed. Careful designs can add greatly to the precision of separation and combination; precise framing of approach and perspective and creative and precise routing can enable experience and attachment while preserving the most vulnerable ecologies. In addition, we could be more careful in our disturbance of complicated and intricate systems that have been around for a very long time. We pollute our lands and, given time lags in these systems, the consequences are often borne by future generations. Lakes and their watersheds are complex systems, and the time lags from watershed pollution are dependent on such factors as lake depth, flushing rate, and the scale of the disturbance. For many lakes the science reassures us that lake restoration is possible, but often at high cost. It is better to protect our lakes with environmentally sensitive designs and good land-use management that avoids substantial alteration of the hydrologic and phosphorus cycle within the lakeshed.

Within this philosophy of community design, both place and time should be considered in the context of nature and landscape. We believe that the landscapes surrounding the places we live should be *open to interpretation.* Our recommendations for a reinvention of lakeshore living only make sense when they allow us to experience the place in multiple ways. Siting, routing, preserving, and clustering all contribute to a new experience of place in context-specific designs, but must still allow for a variety of interpretations of the place by different people at different times. This increases the importance of a fine balance between separation and combination of uses and users of a place. It also reinforces the notion that different networks at different speeds should be combined and subtly interlinked. Moving slowly through a place and observing the various rhythms of nature and the changing seasons helps shift our perceptions of the place. It helps in creating jumping-off places and in opening up new horizons. A new sense of place can emerge more easily when time is experienced differently. Knowledge can help. It seems impossible not to appreciate century-old lichen on billion-year-old boulders.

Rethinking our lakeshore living, then, is offering reinterpretations of old environments and offering new environments that are capable of absorbing and inspiring a variety of meanings, experiences, and uses. Sustainability is a fragile and bounded human endeavor, and the same holds true for justice. Not only are both sustainability and justice continuously renegotiated in shifting political and cultural landscapes; not only is an integration of human uses in unpredictable ecosystems limited; it is also that a utopian understanding of the potential harmony between humans and the rest of nature can be oppressive in its own right. Many people will never agree with Leopold's land ethic, be interested in Olson's wilderness spirituality, or fully understand Whyte's anthropologist-designed city places. Some people do not consider animals, plants, ecosystems, or less prosperous segments of the human population their equals, and the long term is not necessarily relevant for others. The inclusive and long-term perspective of just and sustainable design presented here cannot be assumed to be shared and cannot simply become shared by clear communication, education, and consensus building.

The limitations of consensus formation on just and sustainable design, on the flexibility of humans and the rest of nature, and on the openness to reinvention, reinterpretation, and mutual adaptation are real. The ecological, political, financial, legal, and cultural contexts determine the combinations of human uses possible along a lakeshore. The combinations are limited in time and space and carry their own risks. Knowing a place, an ecosystem, and a community reemerges as a basic condition to identify the scope of combinations in site-specific designs. Identity of place then emerges as part of the same renegotiation marking justice and sustainability. Place identity might be in the eye of the beholder, but it is possible to slowly work on creating communities of observers that savor the uniqueness of a place, as it is possible to capitalize on history and ecology to build places that can be appreciated for their character. The market can be a double-edged sword, a builder and destroyer, and this holds true in the efforts to reinvent lakeshore living. Markets can destroy ecosystems by steamrolling landscapes and imposing uniformity, but they can also create demand for a variety of places and lifestyles. Market values reflect existing desires and desires created by the market. Good lakeshore designs inspired by literature can do the same.

The lessons learned from the lives of Leopold, Olson, and Whyte are manifold, and their projects are not finished. Many lakes have been developed in deplorable ways, and reaching back to their writings can be helpful in many ways. We identified the importance of site-specific, multifunctional, and sustainable design. Within this concept of multifunctionality, ecology and scenery are valuable by themselves. We underlined the importance of thoughtful and thorough design to create beautiful places. We came to understand that knowledge of the local landscape, as well as knowledge of local politics and culture, is necessary to undertake that kind of community design. Good site designs should be informed by the same principles as good local comprehensive planning and good regional planning. Designs that might help in reinventing lakeshore living cannot be conceived in isolation; they must be embedded in community-level designs and regional planning based on sustainable principles and consistently enforced. If they are not, then rules, concepts, or enforcement strategies at different levels will neutralize each other. If they are not, then the complex hydrology of most lakescapes will bring the effects of one flawed policy to other places, to other departments in government, to other aspects of life. Water does not recognize social or political boundaries. Striving for a new type of lakeshore community and for a new way of lakeshore living will require a long-term effort to improve and harmonize planning and environmental regulation efforts on all levels.

Never-ending conservation efforts are an essential part of the proposed design and policy solutions. If place makes people, the precept is that a richer quality of life occurs when we allow and insert more nature into our places. To enhance place identity and accommodate people and the rest of nature, we can acquire some shorelands for parks. Lakeshore residents can restore their shorelines and enrich the visual qualities of place. To promote urban vitality and protect sensitive areas, governments can resume the responsibility of designating the layouts for all future streets and roads. Over time shoreland subdivisions can be reconfigured. Roads can be converted to streets or removed to mend neighborhood streets and communities back together. Streets can be realigned and narrowed with reconstruction opportunities.

Many improvements remain possible after imperfect initial development. The previous chapters gave examples, principles, and strategies, and it bears repetition that one of the main guiding principles of this book was that development needs to be conceived as redevelopment. This simple shift in perspective can drive home the point that context is everything: ecological, cultural, and policy contexts *already exist* and ought to inform and shape what happens next. Our misguided infrastructure and development are not forever. Deepening our land stewardship ethics will give us the confidence to convince our neighbors about the wisdom of good lakeshed and shoreland conservation practices. We will ask our governments to facilitate effective lakeshed planning and large-scale community design. We will create lakeshore places for recreation and we will protect and restore lakeshore areas of wildness for our children and the creatures that depend on these places.

It should now be clear that we are not dealing with a luxury problem affecting only the leisure class and lake lovers, and that we are dealing with a web of issues that cannot be ignored. Failing to design and to leverage the system change that we identified will jeopardize existing lakeshore qualities and values, and we may miss the opportunities to create new qualities for lakes. These lakes are our lakes, and it is our responsibility to create better lakeshore places.

Notes and Recommended Reading

Topics are generally presented in the section or chapter in which they first occur.

PREFACE

Systems Thinking: Senge et al. (1994) provided a good guide to using this approach within an organization, and Meadows (2008) provided a summary of system analysis and how to influence complex systems to obtain desired conditions. One of the first steps in systems thinking requires us to think deeply about a system by writing down our mental models of how we believe the system functions. Important variables influencing water quality or enduring lakeshore communities are often overlooked or can be too difficult to measure. Developing simple conceptual models by diagramming complex interrelationships between variables is helpful when trying to understand lakeshore systems. Selecting variables to include in models is difficult, and the appropriate level of model complexity is dependent on purpose and ease of testing. Human behavior and dynamics are inherently difficult to incorporate, so developing a model framework must first occur.

One approach to begin formulating models for complex systems is the creation of cognitive maps. A cognitive map is a simplistic model of causal relationships among variables. Cognitive maps reduce analysis to a matter of identifying variables, the links among them, and the strength of the links. A cognitive map draws a causal picture, and this simple qualitative approach can synthesize expert knowledge and research findings to predict how complex events interact. Cognitive maps have been used in political science to model political situations, and they are used to model ecosystems. Model complexity is under our control, and models are always simpler than reality. Given this simplification, reality may surprise us, and our predictions may prove incorrect. In addition, a cognitive map may not predict lag effects or how delays in the system change outcomes, which are important in modeling economic systems. However, when one is struggling to identify how to make good things happen, it often helps as a first step to draw out the system by identifying important factors and linkages with a cognitive map. Linkages are either simple relationships between factors (positive or negative) or complex feedback loops within factors (positive or negative). In both the real world and the virtual modeling world, changing the linkages changes the behavior of the system. The act of writing down our view of a system forces us to think holistically about factors and linkages that likely exist or should exist.

With a working model in hand, the next step in understanding the system's behavior is to look at long-term data series, if available. Are there observable patterns in the data, and does the model produce similar patterns? Focusing on the long-term patterns versus the short-term variability often helps us understand the system. For example, it is common for those collecting Secchi disk water-transparency data to focus on the short-term variability in the data. However, these short-term patterns are often the result of differences in annual precipitation. Generally, there is less runoff and nutrient loading to a lake in dry years compared to wet years, and a series of dry years often has greater water transparency than a series of wet years. While this variability is interesting, the greater need is to understand the long-term pattern in water transparency. Are there any consequences of decades of nearshore sewage treatment, agricultural practices within the watershed, or lakeshore urban runoff? For many lakes, long-term data may be hard find or not present, and caution should be taken about inferring the presence of significant trends without these data.

Meadows (2007) spoke of leverage points of change—defined as critical points to intervene in a system. She identified twelve such leverage points and ranked them on their ability to transform a system. Meadows (2008) wrote: "I have come up with no quick or easy formulas for finding leverage points in complex and dynamic systems. Give me a few months or years and I'll figure it out. And I know from bitter experience that, because they are so counterintuitive, when I do discover a system's leverage points, hardly anybody will believe me. Very frustrating—especially for those of us who yearn not just to understand complex systems, but to make the world work better."

Managing Lakes as Part of the Landscape: When managing many lakes, Soranno et al. (2008, 2010) advocated that governments rely on predictive models that use landscape variables (e.g., land use, soils, geology, runoff) and lake classification schemes to set lake conservation goals and develop nutrient criteria thresholds.

Sustainability: In 1983, the United Nations convened the Brundtland Commission (formally called the World Commission on Environment and Development, but so named for its chair, Gro Harlem Brundtland) to address policies for sustainability. The Brundtland Report, titled *Our Common Future* (World Commission on Environment and Development 1987) defined sustainable development as "development that meets the needs of the present without compromising the ability of future generations to meet their own needs." More recently Ehrenfeld (2008) defined sustainability as "the possibility that human and other life will flourish on the Earth forever." Foley (2010) noted that we should focus on the large issues affecting sustainability (e.g., expanding agriculture and irrigation), not the small, distracting issues (e.g., bottled water).

Middle Landscapes: Marx (1964) used the term to address America's countryside ideal, which was being transformed by machines, and the tensions created with this condition. Suburbia was the consequences of this ideal, which was only realized with the advent of the automobile.

PROLOGUE

Trends in Development: Lakeshore development trends were described by Radeloff et al. (2001) for rural areas in the United States and Brown et al. (2005) for northern Wisconsin. Bernthal and Jones (1997) summarized the history of lakeshore development in Wisconsin. They noted that as early as the late 1800s lakeshore resorts catered to wealthy residents of Milwaukee and Chicago, and that the 1950s and 1960s were a period of rapid cabin development for lakes in the northern part of the state. This lakeshore-building boom resulted in Wisconsin enacting one of the first shoreland zoning programs in the United States. Other states, like Minnesota, soon followed Wisconsin's lead.

There are numerous regional studies on the development of shoreland. Haines et al. (2011) studied parcel and habitat fragmentation in Bayfield County, Wisconsin. They found high rates of change from forest to developed land in areas in which parcelization had already begun (e.g., around lakes) and forest fragmentation to be increasing (i.e., parcelization leading to habitat fragmentation). Chi and Marcouiller (2012) found that the availability of public land was an important amenity that attracted people to migrate to northern Wisconsin. Cohen and Stinchfield (1984), Minnesota DNR (1989), and Kelly and Stinchfield (1998) studied changes in Minnesota lakeshore development. Payton and Fulton (2004) estimated that there were about 181,000 lake homes on fish lakes in the state in 2004. About half of all lakeshore homes were seasonal residences, and 75 percent were located on less than two hundred feet of lakeshore frontage (median lot width was 130 feet). The Minnesota DNR estimate for total lakeshore dwellings in 2004 was about 225,000 for all lakes in the state. Development around north-central Minnesota lakes, as indexed by dock sites per mile from DNR aerial photos, has varied by shoreland development class (Radomski 2006). General development lakes have had a faster rate of development than recreational development lakes, whereas natural environment lakes were just beginning to be developed in the 1990s. In 2003, mean development density was 4.0 homes per mile for natural development lakes, 11.2 homes per mile for recreational development lakes, and 18.5 homes per mile for general development lakes. Jakes et al. (2003) modeled future development potential for Itasca County (Minnesota) lakes by identifying seven constructs influencing lakeshore development: current general development, current housing development, availability, accessibility, suitability, aesthetics, and proximity to services. Also in Itasca County, Mundell et al. (2010) studied parcelization rates of the county's forty-acre private lands from 1999 to 2006. They found no trends in the rate and that parcelization was most common adjacent to water and public lands. They noted that parcelization led to development. The Minnesota State Demographic Center has projected growth in many of the lake-rich counties to exceed 35 percent in the next twenty-five years. The central Minnesota lakes area around Brainerd was one of the nation's fasting growing micropolitans (fourth fastest growing mini metro area in the Midwest and twenty-eighth nationally; U.S. Census Bureau 2005).

Effects of Lakeshore Development on People's Perceptions: Studies by Stedman (2003) and Stedman and Hammer (2006) found that attitudes and perceptions about environmental quality changed with changes in density and development, regardless if water quality changed. A Minnesota study found that 27 to 43 percent of people surveyed responded that fishing, scenic quality, water quality, and condition of shoreline on their most-visited lake was "fair or poor," and 85 percent cited development as a cause of decline in scenic quality (Anderson et al. 1999). Respondents reported by a 2:1 margin that lake environments were becoming "worse" rather than "better." A survey of lake associations conducted at the University of Minnesota found that more than 50 percent of respondents felt that water quality, zoning, lake levels, agriculture, exotic species, plants, and fishing were "very important" problems to their lake associations.

Watershed: A watershed or drainage basin is an extent of land from which water drains downhill into a body of water, and it includes the water conveyors or containers (streams, rivers, lakes, aquifers, etc.) as well as the land surfaces in this defined area.

Lawn to Lake: Typical lawn management behaviors lead to high nutrient runoff (Baker et al. 2008b). Garn (2002) found that lakeshore lawn runoff occurred in over half the summer storms, and phosphorus concentration of the lawn runoff was directly related to lawn soil concentrations. In residential areas around lakes, Waschbusch et al. (1999) found the largest phosphorus source was runoff from lawns and impervious surfaces. There are approximately 225,000 residential lake lots in Minnesota, and more than 25 percent have self-reported that they mow a lawn down to the lake (Payton and Fulton 2004). This increases shoreline erosion, eliminates nearshore habitat, and allows unfettered runoff to reach the lake.

Several studies have estimated the consequences of lawn runoff to lakes. Hunt et al. (2006) noted that nitrate and total phosphorus concentrations were three to four times higher in groundwater under lawn areas than wooded areas. The estimate that a lawn-to-lake shoreline allows seven to nine times more phosphorus to enter the lake than a more undomesticated vegetated shoreline is based on the work of Dennis (1986), Bernthal (1997), Bernthal and Jones (1997), and Graczky et al. (2003). Graczky and Greb (2006) studied the interaction of runoff and shoreline buffers and concluded that greater structure setbacks with native vegetation buffers would increase the likelihood that high-intensity rainfalls would be infiltrated before entering the lake. Structure setbacks and shoreline buffers are important because together they enhance a site's ability to absorb water before it is conveyed to public waters.

Agriculture: Vitousek et al. (1986) made the first thorough estimate of human appropriation of global net primary production (HANPP). Haberl et al. (2010) reviewed this topic and summarized other estimates of HANPP. In the most recent estimate humans appropriate about 24 percent of the total primary production of Earth (Haberl et al. 2007), which leaves about 76 percent for the millions of other species. Foley et al. (2007) asked how much is too much ("30%? 40%? 50%? More?") and when will our systems break down. Unfortunately, it is not possible to answer these questions until after catastrophe occurs. Blann et al. (2009) reviewed studies on surface runoff and subsurface drainage from agricultural lands and found evidence for substantial negative consequences of altered hydrology on aquatic ecosystems. Increased subsurface drainage by tiling in Midwest watersheds likely has resulted in an increase in the stream flow–to–precipitation ratio and greater stream flow volumes, specifically in the fall and winter (Lenhart et al. 2011). The first of the last seven of the U.S. EPA's national water-quality reports to Congress was 1992 (U.S. EPA 1994). Agricultural effects on lake water quality are not unique to the United States; for example in Danish lakes, Nielsen et al. (2012) described a positive relationship in the proportion of the lakeshed in agriculture to lake phosphorus concentration.

Eutrophication and Phosphorus Loading Studies: Carpenter et al. (1998) reviewed scientific studies on eutrophication and recommended decreasing nonpoint phosphorus and nitrogen pollution of surface waters. Genkai-Kato and Carpenter (2005) modeled the role of phosphorus recycling in lake eutrophication. An interesting relationship was found for one Wisconsin lake (Gergel et al. 2004): as wealth increased in the drainage basin, water clarity decreased and phosphorus concentrations increased. This model could predict the future for many lakes, as lakes are sentinel ecosystems for regional change (Carpenter et al. 2007). Nearshore algal production can be an early indicator of eutrophication (Lambert et al. 2008). Schindler (2006) discusses the scientific history of eutrophication studies and future challenges, and Schindler et al. (2008) summarize the results of a long-term study demonstrating the importance of decreasing phosphorus loading to reduce eutrophication.

Several studies have estimated sources of phosphorus loading for lakes. Carpenter (2005) simulated lake eutrophication in an agricultural watershed, and results suggested continual inputs of phosphorus from overfertilized soils would inhibit lake restoration. Schussler et al. (2007) and Baker et al. (2008a) in a study of eleven Minnesota lakes using watershed phosphorus balance equations found a range in lakeshed phosphorus retention with several lakes retaining large percentages of phosphorus in their watersheds, which suggested declining water clarity in several to be the result of sewage systems exceeding the soil phosphorus adsorption capacity or agricultural runoff. Runoff from lawns and impervious surfaces can also be large sources of phosphorus pollution. For a southern Wisconsin lake, Garn et al. (1996) determined that runoff, while only a small proportion of the water entering the lake, was the largest source of phosphorus pollution to the lake. For two north central Wisconsin lakes, Garn et al. (2010) estimated that lakeshore runoff and sewer systems accounted for 22 percent of the phosphorus loading to the lakes.

Paleolimnology studies that have reconstructed past water-quality conditions include Garrison and Wakeman (2000) and Garrison et al. (2010) for Wisconsin lakes, Heiskary and Swain (2002) for Minnesota lakes, and Albert et al. (2010) for shallow northeastern lakes. Ramstack et al. (2004) found about one-third of Minneapolis and central Minnesota lakes had a significant increase in total phosphorus between 1800 and the present. These changes were attributed to increases in nutrient runoff from agricultural and urban land. Lakes in the forested region of northeast Minnesota saw little change in lake total phosphorus concentrations.

Need for Rainwater Management: The U.S. EPA (1998) noted that rainwater runoff is an important source of lake pollution, and Minnesota PCA (2001) noted that sediment transport to lakes is a large pollution problem.

Sediment Erosion: Owens et al. (2000) conducted a study of small construction sites and found sediment loads from two monitored construction sites were ten times larger than typical loads from rural and urban land uses in the area.

Impervious Surface: Comprehensive studies or summaries on the negative consequences of failure to manage stormwater from impervious surfaces include Schueler (1994, 2003, 2008), Schueler and Caraco (2001), and Schueler et al (2009). Research has documented how impervious areas can alter stream hydrology (FISRWG 1998). There is a definitive link between impervious surface cover and ecosystem condition. Studies on impervious surface coverage and relationships to fish, wildlife, and habitat health include Wang et al. (2001) and Brabec et al. (2002), which again stresses the importance of large natural vegetated shoreline buffers (DeLuca et al. 2004). Cappiella and Schueler (2001) recommended a 10 to 15 percent impervious cover limit in shorelands to protect water quality. Barten (2005) found that even lawns could act as impervious surfaces if soils were compacted during construction or with heavy use. Extensive investigation and research has shown the link between proportion of watershed in impervious surface cover and degraded water quality (Schueler 2003) and fish assemblages (Wenger et al. 2008). Schueler et al. (2009) reviewed sixty-five scientific studies on impervious surface cover and found that a majority of the research supported the predictive model that as impervious surface coverage increases, stream health decreases due to hydrologic changes, declines in water quality, and loss of habitat.

In addition, increased imperviousness results in increased stream water temperatures and reduced base flows. Increased imperviousness lowers base flow because less precipitation infiltrates into the groundwater, leading to a lowered groundwater table. Increased stream temperatures and high annual temperature fluctuation have a negative impact on fish communities, particularly for fish that thrive at cooler water temperatures. Generally, it has been observed that at between 10 to 12 percent imperviousness in the watershed there is a decline in stream fish communities, and above 25 percent fish are usually absent (Paul and Meyer 2001). Increases in imperviousness also affect species richness. In Wisconsin, fish and insect coldwater indices of biotic integrity decreased rapidly at 10 percent urban land cover (Wang et al. 1997, 2003), and 8 to 12 percent connected impervious surface coverage resulted in major changes in stream condition (Wang et al. 2001; Wang and Kanehl 2003). Wang et al. (2000) also found a threshold of environmental damage to streams at about 10 percent imperviousness and concluded that while agricultural land uses often degrade stream fish communities, their impacts can be less severe than those from urbanization, which results in substantial changes in runoff volume and pollutant loads. In Minnesota and Wisconsin, trout streams degraded quickly at 6 to 11 percent connected impervious surface coverage, indicating that even low levels of urban development can damage these streams. In addition, Cuffney et al. (2010) found that even low levels of urbanization reduced stream aquatic invertebrate populations.

Consequences of Road Deicing on Lakes: Several studies have described trends and the negative impacts of this salt (sodium chloride) use on aquatic ecosystems (Ramstack et al. 2004; Sanzo and Hecnar 2006; Kelly et al. 2008; Novotny et al. 2008; and Palmer et al. 2011).

Habitat Loss: The key finding of the U.S. 2007 National Lakes Assessment, the first nationally consistent assessment of U.S. lakes, was that 44 percent of the nation's lakes are in poor biological condition. In addition, the assessment concluded that poor lakeshore habitat is the biggest problem, followed by high nutrient levels (U.S. EPA 2009).

Wildlife Research on Consequence of Loss of Shoreline Buffers: An early study on loss of shoreline buffers and natural vegetation was conducted in Ontario by Clark and Euler (1984). Elias and Meyer (2003) documented shoreline vegetation and habitat losses along northern Wisconsin lakes, and Lindsay et al. (2002) found changes in the associated bird communities. Henning and Remsburg (2008) found higher bird species richness at lakeshore sites with intact understory vegetation than those without. Bald eagles that nest on developed and altered shores have to spend significantly more time and energy feeding (Fraser et al. 1985). Green frog population reductions due to loss of shoreline buffers were found by Woodford and Meyer (2003). Marczak et al. (2010) reviewed the effectiveness of riparian buffers in providing habitat for wildlife, and they concluded that commonly prescribed buffer widths (15 to 150 feet) seldom provided sufficient wildlife habitat and that much wider buffers that incorporated upland forest areas were necessary for sensitive species.

The loss of wood habitat in lakes has been documented by Christensen et al. (1996), Jennings et al. (2003), Marburg et al. (2006), and Francis and Schindler (2006). The density of nearshore woody habitat was negatively correlated with dwelling density, and developed shorelines had only 15 percent of the average wood habitat (logs/shoreline length) as forested shorelines (Christensen et al. 1996). Heavily developed lakes rarely had much wood habitat (Marburg et al. 2006; Wehrly et al. 2012). Fish and wildlife are negatively affected by the removal of downed wood in the littoral zone (Newbrey et al. 2005b; Roth et al. 2007; and Tabor et al. 2011). For example, turtles are often dependent on deadfalls or floating logs for their basking sites (Boyer 1965). In addition, Francis et al. (2007) concluded that the loss of downed wood alongshore altered the sediments, food webs, and energy flow within lakes.

Shoreline development also influences lake food webs in subtle ways. For example, fish may lose food sources from shore; Francis and Schindler (2009) noted that trout in developed Pacific Northwest lakes had eaten fewer insects originating from the shoreland than trout from undeveloped lakes. They attributed this difference to deforestation of developed lakes (the shoreline vegetation density on developed lakes was approximately one-tenth that of the undeveloped lakes) that resulted in fewer habitats for shoreline insects and trout prey. Using stable isotope analysis, Larson et al. (2011) found similar diet losses for crayfish.

Water-Quality and Other Benefits of Shoreline Buffers: Bentrup (2008) provided a review of the benefits and design guidelines for shoreline buffers organized by water quality, biodiversity, economic opportunities, aesthetics, and other resources. This report referenced over 1,400 publications that were used to develop the guidelines. Other scientific research on shoreline buffers includes Castelle et al. (1992, 1994), Desbonnet et al. (1995), Woodard and Rock (1995), Wenger (1999), Wang et al. (2003), Weigel et al. (2005), Mayer et al. (2007), and Diebel et al. (2009). In comparing strategies to reduce phosphorus loading to lakes, Kramer et al. (2006) suggested that shoreline buffer restoration could be a more cost-effective strategy than sewer upgrades.

Most Minnesotans strongly agreed that the aesthetic value of the state's lakes is important to protect and that they would support regulations that limit human use to protect lake resources (Schroeder et al. 2004). However for aquatic plants, about one-third of surveyed lakeshore property owners remove plants, and most respondents may have been influenced by the perceived disadvantages of plants near their shoreline (Schroeder and Fulton 2013).

In Wisconsin, research by Macbeth (1992) and Engel and Peterson (1998) has shown that people prefer to view lakeshores where the vegetation screens manmade structures.

In addition, lawn-to-lake management is expensive for the lake home owner. For example, businesses save between $270 and $640 per acre in annual mowing and maintenance costs when they convert to natural buffers instead of lawns (Wildlife Habitat Enhancement Council 1992, as cited in Schueler and Holland 2000). A national survey of thirty-six stream buffer program administrators indicated that stream buffers were perceived to have either a neutral or positive impact on property values, and none of the respondents indicated that buffers had a negative impact on land value (Heraty 1993). Lake shoreline buffers are also expected to have positive impacts on property values. These findings and expectations are consistent with other studies that have found that greenways, natural areas, and buffers increase property values (Correl et al. 1978; Anton 2005; Crompton 2007).

Loss of Shoreline Habitat with Development: Changes in shoreline plant communities may also occur with lakeshore development. Developed shorelines had 92 percent less floating-leaf and 83 percent less emergent vegetative cover than undeveloped shorelines for a group of Wisconsin lakes (Meyer et al. 1997). Radomski and Goeman (2001), in comparing undeveloped and developed shoreline plots, estimated a 20 to 28 percent loss of emergent and floating-leaf coverage from human development for all of Minnesota's clear-water, centrarchid-walleye lakes. For developed shoreline plots they found a 66 percent reduction in vegetative cover relative to undeveloped plots. In northern Wisconsin, Jennings et al. (2003) also found that emergent and floating-leaf vegetation decreased at developed sites and in lakes with greater cumulative shoreland development density. Elias and Meyer (2003) found that tree and shrub coverage and the mean number of shoreland plant species were lower along developed shorelines than along undeveloped shorelines. Hatzenbeler et al. (2004) determined that aquatic plant communities declined with increasing lakeshore development, with the frequency of occurrence of floating-leaf vegetation lower on more developed lakes. Hicks and Frost (2011) noted a negative correlation between aquatic plant richness and lakeshore development, and they found species composition differences related to shoreline house density (Frost and Hicks 2012). In addition, quantifying long-term changes in aquatic macrophyte communities is important in the assessment of ecological consequences of human activities. Radomski (2006) found that floating-leaf and emergent vegetation coverage in Minnesota lakes declined over a sixty-year period and that this habitat coverage was significantly affected by development.

Human activities that change vegetative cover can alter ecological processes and energy flow within lakes, thereby reducing their ability to support diverse and healthy fisheries (Bryan and Scarnecchia 1992; Schindler and Scheuerell 2002). For example, aquatic emergent vegetation loss negatively affects black crappie and largemouth bass populations (Reed and Pereira 2009). To understand the potential impacts of docks on lake shoreline habitat, Radomski et al. (2010) conducted an inventory of docks across north-central Minnesota and simulated full build-out projections. Docks were pervasive along many of the lake shorelines, and over 14 percent of the shoreline and 3 percent of the littoral zone were estimated to be impacted by docks. Build-out scenarios estimated that up to half of the shoreline and 14 percent of the littoral zone could be impacted with future development. The investigators concluded that shoreline development policies may need to be revised to address impacts to fish habitat and recreational surface-water use.

Subsurface Sewage Treatment Systems: According to the Minnesota PCA (2004b), 39 percent of individual sewage treatment systems were estimated to be failing or posed an "imminent" threat of creating serious nutrient and bacterial contamination. The agency also found that nitrate concentrations in domestic water wells increased with age of sewer systems in the area (Minnesota PCA 2000) and that nitrates exceeded drinking-water standards down-gradient of sewer systems (Minnesota PCA 1999). Steffy and Kilham (2004) found elevated nitrogen-15 in macroalgae, aquatic insects, crayfish, and small fish collected near subsurface sewage system areas compared to samples collected in unimpacted sites, indicating that nitrogen pollution from sewage systems had reached surface waters. In deepwater lake embayments and nearshore areas of many lakes, nitrogen pollution originating from lakeshore sewer systems can lead to fertilization of aquatic plants (e.g., Benson et al. 2008).

Numerous studies have shown that phosphorus movement toward the lake from sewer system drain fields can be substantial. The capacity of the soil to retain phosphorus is finite, and phosphorus movement deeper into the soil profile and down-gradient to water resources can be expected (Minnesota PCA 2004c). Minnesota PCA (1999) found elevated phosphorus concentrations in groundwater within fifty feet of fully functioning sewer systems. Dillon et al. (1995) determined that phosphorus from subsurface sewer systems might have been an important pollutant source for four Ontario lakes. Several studies have found that phosphorus from sewer systems could reach surface waters due to impervious layers, such as bedrock, beneath surface soils, which could be an issue for northern lakes (Ptacek 1998).

Robertson (2003) found a distinct phosphate plume that extended about ninety-eight feet down-gradient of the drain field at a calcareous sand soil site. Long-term monitoring of phosphate levels of sewer system plumes on calcareous sand showed that after seventeen years a distinct phosphate plume was present and it extended sixty-six feet down-gradient from the drain field. Migration velocity was about three feet per year, reflecting retardation by a factor of twenty compared to the groundwater velocity (Robertson 1995; Robertson et al. 1998). In contrast, noncalcareous sand sites had no phosphate plumes or they were more localized with lower concentrations of phosphorus; also at these sites, phosphate concentrations in the shallow water table zones below drain fields did not increase over time (Robertson 2003). Robertson (2008) also reported on a long-term study of a shoreland subsurface sewage treatment system. Over a sixteen-year period, a phosphorus plume extended more than fifty feet from the absorption area. These investigations show that in calcareous sands with high permeability, phosphorus is not contained within the standard setbacks from lakes and rivers. Thus, phosphorus can remain mobile and pose a serious threat to lake and river water quality in these terrains.

Maintenance of sewer systems is critical. Sludge builds up in the sewer tank and should be pumped out every two to three years. If sludge accumulates to the level of the outlet pipe, clogging will occur. This will damage the drain field, reducing the life expectancy of the system. Drain fields can also fail when they are overloaded, either with too much water or with garbage disposal waste in volumes higher than designed for the system. The average life of a drain field is ten to twenty years. Lake home owner management of sewer systems is sometimes inadequate. Regular pumping of the sewer tank is needed to minimize pollution problems. Some areas have developed comprehensive management programs that track routine maintenance and compliance with public health standards. These programs can save homeowners money, because regular maintenance and inspection costs are much less than replacement of failed systems. Information on advanced sewer systems plus tips on sewer system operation and maintenance can be found online at the University of Minnesota's Water Resources Center at http://septic.umn.edu.

CHAPTER 1. LAKE PARTS

Distribution of Lakes: Downing et al. (2006) described the extent and distribution of lakes on planet Earth.

Benoit Mandlebrot: Dr. Mandlebrot (1924–2010) was a mathematician best known for his advancement of fractal geometry. Fractal geometry, which he named, involves the roughness of an object and is a field of science related to chaos theory and complexity. He wrote *The Fractal Geometry of Nature* in 1982, which introduced the topic to nonscientists and triggered scientific work into chaos theory. His ideas changed the way we look at common things in nature, like clouds, trees, mountains, leaves, and snowflakes. Instead of assuming that complex natural rules exist for these things, he showed and stated that "bottomless wonders spring from simple rules, which are repeated without end."

Aquatic Plants: Nichols and Vennie (1991) and Valley et al. (2004) discussed the attributes and value of aquatic plants. Good aquatic plant identification books include *Aquatic and Wetland Plants of Northeastern North America,* volumes 1 and 2 by Crow and Hellquist (2000), *A Manual of Aquatic Plants* by Fassett (1957), and *Aquatic Plants of the Upper Midwest* by Skawinski (2011). Water chemistry strongly influences the plant species that occur in a lake. In Minnesota lakes, Moyle (1945, 1956) described the influence of alkalinity on the species of aquatic plants found, and Radomski and Perleberg (2012) found that the aquatic plant richness decreased with increasing phosphorus concentration and lakeshed disturbance.

Dragonflies and Damselflies: Two good sources of information on these insects are *Dragonflies of the North Woods* by Kurt Mead (2009) and *Damselflies of the North Woods* by Robert Dubois (2005). Remsburg and Turner (2009) found higher adult damselfly abundance in areas with more abundant shoreline plants. The emergence of aquatic insects from lakes is an important food source for terrestrial predators (Gratton and Vander Zanden 2009; Vander Zanden and Gratton 2011).

Fish, Fish Movement, and Habitat Use: *Fishes of Wisconsin* by George Becker (1983) is a classic treatise on fish found in northern lakes and rivers. *The Great Minnesota Fish Book* written by Tom Dickson and illustrated by Joseph Tomelleri (2008) provides a reference to learn about both familiar and unfamiliar fish. Landsman et al. (2011) summarize fish movement studies in the Great Lakes.

Sensitive Fish Species: Pugnose shiner, least darter, longear sunfish, blackchin shiner, blacknose shiner, and banded killifish need areas within the lake that have extensive aquatic vegetation (Valley et al. 2010), and they have been extirpated from lakes where watershed and lakeshore development have occurred (Clady 1976; Lyons 1989).

Northern Pike: Pierce (2012) described the life history and management of northern pike.

Loons: McIntyre (1988) and Evers et al. (2010) described common loon biology. Mercury pollution emanating from coal burning has reduced loon reproduction in Maine and New Hampshire (Evers et al. 2008). Loons ingest lead fishing tackle when they mistake small lead fishing sinkers and jigheads for the small stones they pick up from the lake bottom to help them digest their food or when they ingest fishing line with a lead sinker still attached to a baited hook. A single ingested lead sinker or jighead will kill a loon. The use of nonlead fishing tackle would reduce loon deaths (Scheuhammer 2009). Loons appear to adapt to human activity; Titus and VanDruff (1981) found that loons nesting on high recreational use lakes flushed from their nests less often than those that nested on low recreational use lakes. However, loons appear vulnerable to shoreline development. Loon nesting studies showing that the probability of loons on the lake decreases with increased housing density include Newbrey et al. (2005a) and Caron and Robinson (1994). Radomski et al. (2013) found that shoreline areas with fewer developed parcels had higher probabilities of loon nesting.

Humans: The carrying capacity of a species is the limit set to its population size by the availability of resources (Pulliam and Haddad 1994). Carrying capacity is dynamic, as ecosystem conditions are dynamic. Carrying capacity can be exceeded, but only temporarily. With regard to humans, carrying capacity is also contingent on our technology and culture. Have we reduced or eliminated species that we compete with or exploit? We caused the extinction of the passenger pigeon with our exploitation, numerous mussel species with our water pollution, and perhaps the Neanderthals with our competition for resources. Have we expanded in numbers to a level sustainable only with ample supplies of oil or a suite of high-energy sources?

While some people, including some economists, assume that our population growth has no biological limit, most ecologists do not hold the same view. There is always a limit. We can judge how hard we are pushing the environment with our population and activities by observing the trends in ecosystem resiliency, i.e., the capacity of the system to absorb disturbances without change in structure, function, and character. Are we losing soil and agricultural productivity? Are systems downstream of our cities and croplands dramatically changing due to runoff? Is biodiversity diminished such that food webs are substantially altered? Do we see accelerating lake fertilization (eutrophication)?

Arrow et al. (1995) generated considerable discussion in the scientific community with an article on economic and environmental policy in the face of increasing demands on natural resources due to human populations (e.g., see Ludwig 1996 and Daily et al. 1996). Jared Diamond's books on human societies, *Guns, Germs, and Steel* (1999) and *Collapse* (2005), are interesting readings on the importance of the geographic distribution of natural resources and the mechanisms for human population collapse. Vitousek et al. (1997) and Foley et al. (2007) discuss the large transformation of the planet that has occurred to support our large populations. Hurlbert (2012) expressed his opinion about the Ecological Society of America's difficulty in addressing human population growth in the context of becoming sustainable, and he believed that even an organization of ecologists might have found human population growth policies a taboo subject.

CHAPTER 2. LAKE ECOLOGY

Principles of Ecology: The four principles presented are based on the seven principles that Knapp and D'Avanzo (2010) identified. Barry Commoner's laws of ecology are perhaps more widely known and cited. In 1971 Commoner wrote in *The Closing Circle* the following four laws:

1. Everything is connected to everything else. The single fact that an ecosystem consists of multiple interconnected parts, which act on one another, has some surprising consequences.
2. Everything must go somewhere. In nature there is no such thing as waste. Nothing "goes away"; it is simply transferred from place to place.
3. Nature knows best. Any major man-made change in a natural system is likely to be detrimental to that system.
4. There is no such thing as a free lunch. In ecology, as in economics, the law is intended to warn that every gain is won at some cost.

Limnology: A good scientific treatise on the study of lakes is *Limnology: Lake and River Ecosystems* written by Robert Wetzel (2001).

Sulfur Oxides in Lakes (Acid Rain): Acid precipitation remains a serious problem for many northern lakes with low acid-buffering capacity (Clair et al. 2011).

Carbon Dioxide: Many lakes with nearly neutral surface water (hydrogen ion concentration as expressed in terms of pH of around 7) are slightly supersaturated with carbon dioxide relative to the atmospheric pressure of carbon dioxide already, and therefore export carbon dioxide to the atmosphere. Oceans, however, absorb carbon dioxide. With global warming, oceans will stratify more, resulting in less deepwater mixing, which is where carbon is stored. Surface ocean water will accumulate carbon dioxide, which may lead to the acidification of the world's oceans (carbon dioxide plus water creates carbonic acid, a weak acid).

Phosphorus as a Limiting Factor: Carpenter (2008) reviewed the literature on phosphorus and nitrogen in the role of lake algal production and found that phosphorus control is necessary to decrease eutrophication.

Shallow Lakes: Evidence of degradation of shallow lakes has been documented by many natural resource management agencies (e.g., Minnesota PCA 2004a). The theory of two alternating stable states of shallow lakes has been well documented (Scheffer et al. 1993; Scheffer and Carpenter 2003; Scheffer and van Nes 2007; and Hobbs et al. 2012).

Waterfowl production is reduced with high density development of shallow lakes. Disturbance can cause female nesting ducks to take flight, leaving eggs exposed and chicks more prone to predation. Korschgen and Dahlgren (1992) reviewed over two hundred journal articles that revealed that disturbance from development displaced waterfowl from feeding grounds, increased energetic costs associated with flight, and likely lowered productivity of nesting or brooding waterfowl. Kahl (1991) found that disturbance in a Wisconsin lake resulted in about a 50 percent reduction in feeding time for canvasbacks. Knapton et al. (2000) found that disturbance leads canvasbacks, redheads, and scaup to feed in less productive areas. Belanger and Bedard (1990) found for snow geese that disturbance caused a 5.3 percent increase in hourly energy expenditure.

Asplund (1997) studied water clarity for both weekdays and weekends for shallow and deep lakes in Wisconsin. Boat density increased on weekends, and water clarity decreased by about sixteen inches in the shallow lakes and about eight inches in the nearshore areas of all lakes. Beachler and Hill (2003) found that at boat speeds near six to eight miles per hour, when boats were near-plane, there was maximum turbulence at the lake bottom in shallow areas (less than eight feet deep). Resuspension of lake sediments was less at high or idle speeds. Boat traffic on shallow lakes can result in an increase in phosphorus concentrations (Yousef et al. 1980). This phosphorus can then stimulate growth of attached or planktonic algae, thereby degrading or eliminating important aquatic plant communities (Murphy and Eaton 1983). In addition, boat traffic on shallow lakes and in littoral areas can damage or destroy aquatic macrophytes (Asplund and Cook 1997, 1999; Asplund 2000).

Long-term Changes in Lake Ice-Cover and Climate Change: Around the world, lakes are experiencing longer periods of open water (Magnuson et al. 2000; Magnuson 2002). Climate change will alter lake habitats and many fish communities (Stefan et al. 1996; Schindler 2001; Rahel 2002b; Kling et al. 2003; Mackenzie-Grieve and Post 2006; Ficke et al. 2007). Sunfish and bass will expand their range north, walleye will decline at their southern range and increase at the northern end of their range, and coldwater species such as burbot, lake trout, lake whitefish, and cisco are likely to increase in parts of the Great Lakes, decline in small lakes, and be eliminated in many lakes at the southern end of their range. Research on likely cisco declines due to global warming include Jacobson et al. (2010), Fang et al. (2012), and Jiang et al. (2012).

Sensitive Lakeshore Identification: Several natural resource agencies have programs to identify sensitive lakeshore or critical lake habitat. Integrating natural resource information is often the basis for decision making. Wisconsin designates critical lake habitat based on the quality of the resource, existing scientific data, and potential risks from riparian development and lake recreation activities. In addition, the natural resource management agency considers public rights features on the lake before producing maps and supporting documentation for public review. After public comments are addressed, critical areas are designated. Elsewhere ecological models are often used to assist in the determination of sensitive areas. For example,

Minnesota DNR (2011) uses an approach that is founded on the ecologically based guidelines for land use (Dale et al. 2000; Zipperer et al. 2000) and techniques to identify important green infrastructure (Benedict and McMahon 2006). Two models are used. First, an ecological model that incorporates documented lakeshore plant and animal communities and hydrological conditions is used to identify sensitive lakeshore. The benefit of this approach is that criteria come from science-based surveys (variables include species presence, biological diversity, and habitat size and quality), and the value (or model score) of the shoreline with regard to fish and wildlife habitat is objectively assessed. Second, predictive models are used to identify lakeshore where plant and animals surveys are not available. These models use statistical algorithms, such as logistic regressions, or spatial analyses on hydrological, morphological, and aquatic vegetation variables.

Critical habitat is often clustered, so identification and protection of sensitive lakeshore will have benefits to fish and wildlife. Protected bays often possess a large proportion of the valuable floating-leaf and emergent plant habitat for a lake (Radomski 2006). Numerous fish species use protected embayments and vegetative cover disproportionately to their availability (Wei et al. 2004), and fish densities are often highest in areas of diverse habitat (Sharma et al. 2012). Francl and Schnell (2002) suggest some areas of lower development density within shorelands may maintain regional diversity of birds and plants. Loons would benefit if critical nesting areas were reserved, protected, or had lower lakeshore development densities (Vermeer 1973; Robertson and Flood 1980; Heimberger et al. 1983). Other waterbirds need aquatic vegetation and wetlands for nesting and feeding areas. Finally, these areas are important fish spawning and nursery areas.

INTRODUCTION TO PART 2. SCIENTIST, WRITER, AND ACTIVIST

Science: For an excellent analysis on how science progresses, read *The Structure of Scientific Revolutions* by Thomas Kuhn (1996). Kuhn describes both the accumulated, linear advance of reliable knowledge and, more importantly, the paradigm shifts that occur when scientists or the scientific method produce new ideas or ways of interpreting the world around us. Burnham and Anderson (2002) and Anderson (2007) provide a well-grounded and statistically valid approach to acquire reliable knowledge. Today, model-based inference is the state-of-the-art approach to data analysis. A suite of models is judged by Akaike information criteria (AIC; or, in the Bayesian context, deviance information criteria). The basic idea behind AIC is that models get penalized for complexity (i.e., the number of explanatory variables used in the model). The approach has considerable merit, and it has become the cornerstone of judging predictive models. Akaike weights, then, can be used to quantify the strength of evidence for alternative models.

Poems by Aldo Leopold: Leopold wrote poetically about the extinction of the passenger pigeon ("On a Monument to the Pigeon," *A Sand County Almanac*) and sounds he heard in the southwest river valley—sounds that are not audible to all ("Song of the Gavilan," *A Sand County Almanac*). In the immediate aftermath of the Japanese attack on Pearl Harbor, Aldo Leopold wrote a poem titled "Yet Come June" (unfinished manuscript, dated December 23, 1941) that spoke of the decline in society and an environment substantially degraded.

CHAPTER 3. ALDO LEOPOLD AND LIVING IN HARMONY WITH THE LAND

Posthumous Collections of Essays and Writings by Aldo Leopold: Those publications compiled from Aldo Leopold's archives include Flader and Callicott (1991), Brown and Carmony (1995), Callicott and Freyfogle (1999), Meine and Knight (1999), and Leopold (2013). The Aldo Leopold Foundation, in Baraboo, Wisconsin, is the steward for the material in the Aldo Leopold Archives housed within the University of Wisconsin Digital Collection Center.

Aldo Leopold's Life: The biographical information about Aldo Leopold was based on the excellent biography written by Meine (1988).

Mississippi River Backwater Lakes: Aldo Leopold's words about the loss of the river bottoms and backwater lakes were from the draft forward to the "Great Possessions" manuscript (*A Sand County Almanac*), printed in Callicott (1987). Grubaugh and Anderson (1988) studied one hundred years of water levels and the hydrologic condition data from Pool 19 in Iowa, and they found that floodplain habitat loss was substantial. Similar studies have documented considerable changes to the Upper Mississippi River backwaters. Luna Leopold (1915–2006), son of Aldo and noted hydrologist, studied and wrote extensively on rivers and their floodplains (Leopold 1974, 2006; Leopold et al. 1995).

Tragedy of the Commons: Hardin (1968) popularized the concept. Daily et al. (1997) defined ecosystem services. For more on the tragedy of ecosystem services, see Ruhl et al. (2007) and Lant et al. (2008).

Lake Mendota and the Yahara River Watershed: Recent research on Lake Mendota concluded that a 50 percent reduction in total phosphorus loading from the watershed was needed to substantially improve water clarity. How did researchers estimate this? They had a long time-series of water-quality data, a hydrological system model, and a lake model. It was noted that the drought of 1987–1988 had better water clarity than the wet period of 2002–2004. So they used the drought conditions and understanding of lake phosphorus dynamics to provide a useful total phosphorus reduction target (Lathrop 2007; Carpenter and Lathrop 2008).

Ernest Oberholtzer: Ernest Oberholtzer (1884–1977) spent most of his adult life in a cabin on Rainy Lake, Minnesota. He started with a single small cabin in 1920 perched on a little island he called Mallard, 1,200 feet long and very narrow. By 1950 he had built eight more cabins, although only one cabin was really weatherproof and only one other structure could be labeled a house. Oberholtzer probably never made a design plan for the island. He added things when he found time and money and when the other places where getting overcrowded with books and visitors. Still, Oberholtzer, who studied landscape architecture at Harvard, was very sensitive to the possibilities and fragilities of the site. He took great care to avoid pollution and went to great lengths to find the best sites for every house, the most appropriate ones for the activities he had in mind. His Japanese House, a study with screened-in porch, referred to Japanese rural architecture while fitting in perfectly in the northern landscape, precariously balancing over the water on a rocky outcrop. All nine cabins blended into the natural background and harmonized with each other in style. Oberholtzer did nothing haphazardly, and the lack of a paper design plan obscures the presence of his vision for the place, albeit one that was constantly evolving (Van Assche 2013).

Oberholtzer became famous for his epic battle against lumber baron Edward W. Backus (1860–1934) for the preservation of the Boundary Waters Canoe Area Wilderness. He is rightfully seen as the pivotal figure in the protection of the Boundary Waters and of Voyageurs National Park and one of the leaders in the American wilderness movement. In 1927, Ernest Oberholtzer and his circle of friends/allies realized that they had to do more than just oppose the plan of Backus to turn the Rainy Lake and Lake of the Woods watersheds, on both sides of the border, into a monofunctional water reservoir, with the single aim to generate electricity. In Backus's plan, a series of dams had to be built, and water levels would be raised dramatically. Backus already had several dams in place, and these had caused significant flooding, with damage to properties and to nature. The wilderness character of the area was already affected. Oberholtzer and others opposed the plan, studied its details and implications, informed the public, and lobbied with politicians and courts, but still found themselves inadequate. In the end, they won the battle, but it took decades to do so. In retrospect, two additional actions were critical for Oberholtzer's victory. He and his allies created a counterproposal, a plan for the whole area that would preserve its wilderness character while allowing for use and development in specified ways. The plan was regional, site-specific, multifunctional, and respectful of the spirit of the place. The plan made it possible to combine the interests of many, and it was politically and rhetorically important, since it turned the early preservation movement into a group with a positive message.

The second thing that was needed was a change in mentality. That took a long time. Although Oberholtzer was working for the public interest, it took time before it was seen as such by the wider public and the politicians. It took inspiration and initiative from people like Oberholtzer and his friends to define a public interest before the public was interested. This is often the dilemma for planners: Whose values are we protecting? Can we move ahead of the public? This is essentially a political question, not a scientific one. Oberholtzer was eventually able to persuade the public and its representatives that northern Minnesota and southern Ontario were of great value in a natural state. In a pamphlet opposing the Backus plan, Oberholtzer wrote, under the heading "Damage to intangible values" that "because the region is largely in a state of nature, Mr. Backus assumes that it is valueless; but that is exactly what constitutes its value for the public."

Later, he said that in order to develop his plan for the region, he "went back over all I knew from my years up here [on Rainy Lake], my college studies in landscape architecture and my knowledge of the forest service." He delved into his own extensive local knowledge and his understanding of the landscape, the ecology, water, and visual qualities; he used his landscape architecture and site-planning skills, acquired at Harvard many years ago, and he tried to use his insights into politics and the bureaucracy. All these things together enabled him to come up with a new vision for the whole region. In this process of regional planning, the scale of the region was not always important. Many principles apply at several scales. Joe Paddock, in his beautiful biography of Oberholtzer, *Keeper of the Wild* (2001), observed: "His single year in graduate school . . . was not the sole source of the plan he was now developing. . . . When talking about the plan, Ober most often returned, not to his more advanced study, but to the undergraduate paper he wrote on the best use of the pond that provided freshwater for Cambridge, Massachusetts, in which his chosen remedy was not simply a wilderness remedy. . . . It was his knowledge of the region and the conflicting interests of the people who lived and worked in it that provided the substance for his solution."

Oberholtzer did not write as much as the other fathers of American environmentalism, but his green credentials are widely recognized. Again, as so often with the complex figure of Oberholtzer, appearances are deceiving. All along the drawn-out battle for the Boundary Waters, he kept an eye on the interests of different user groups. He never forgot that people lived there, needed to make a living, and wanted to recreate, investigate, and contemplate in the vast expanses of forests and lakes. The hermit of the island proved a talented networker, a natural focus of the forces that recognized the exceptional value of Minnesota's natural heritage. A large portion of the Rainy Lake watershed that he wished to see as a world war memorial forest is now protected. People all over the world enjoy Quetico Provincial Park, Voyageurs National Park, and the Boundary Waters Canoe Area Wilderness.

Oberholtzer was born in Iowa but developed a strong connection with Minnesota's lakes, wetlands, and forests. He understood the ecological riches of lakes and wetlands and their vulnerability. In his Harvard years, his study projects showed a preoccupation with multi-functional wetland areas. Even then, he was searching for ways to preserve lakes and wetlands by making them useful for many people and many things. The challenge was, and still is, to integrate those activities with each other, and into the landscape, respecting the ecological and visual qualities of the place. His battle was unfortunately not the last one, and his search is not finished yet. The Boundary Waters may have been preserved, but many other lakes have been developed in ways that minimize ecological and visual qualities.

Mr. Babbitt: This character created by Sinclair Lewis (1885–1951) for his fictional book *Babbitt* made quite a lasting impression on Leopold. Mr. Babbitt must have resonated with Aldo's experiences, as Aldo Leopold would include Mr. Babbitt (or just Babbitt), without reference, in many of his writings. Leopold assumed that readers would know of the character; perhaps he thought this because Sinclair Lewis was the first American to be awarded the Nobel Prize in Literature in large part due to the book *Babbitt,* and the book was made into a silent film in 1924 and a sound film in 1934.

Odysseus: Leopold used an event from *The Odyssey,* a Greek epic poem attributed to Homer, to demonstrate the evolution of ethics. Odysseus, finally returning home after a twenty-year odyssey following the fall of Troy, kills a group of lawless suitors to his wife and hangs a dozen disgraced household maids. Leopold noted the maids were slaves, who were considered property, and the ethics of Greece at that time did not extend to such humans. Today our ethics forbids human slaves, extends to all humans, and includes the humane treatment of some animals. Leopold ("The Land Ethic," *A Sand County Almanac*) said it was now time to extend our ethics to include treatment of "soils, waters, plants, or collectively: the land." Caring for the land is in our best interests, and to foster this caring necessitates the evolution of our ethics to include the land.

Norman C. Fassett: Dr. Fassett (1900–1954) was a colleague of Aldo Leopold at the University of Wisconsin–Madison and the authority on aquatic plants. Norman Fassett was born in Massachusetts and as a boy took extensive excursions in the fields, woods, and shorelines of New England. The Maine coastline and estuaries inspired him to focus on aquatic plants. Fassett received his botany training and degrees from Harvard University. Dr. Fassett was professor of botany at the University of Wisconsin–Madison from 1925 to 1954, and he was the curator of the university's herbarium from 1937 to 1954. He likely sparked Aldo Leopold's interest in the identification and life cycles of flowering plants. Fassett was one of the founding members of the American Society of Plant Taxonomists. He wrote *Spring Flora of Wisconsin* (1931), *Grasses of Wisconsin* (1951), and the classic book on lake and river vegetation, *A Manual of Aquatic Plants* (1940). These books are still used by botanists today, and they still can be purchased. Fassett's book on aquatic plants was the definitive guide for the identification of lake and river plants of eastern North America. (Shortly before his death, Fassett asked a colleague to update the plant nomenclature of the book; this effort resulted in an appendix being added to the book in 1957.) In *A Manual of Aquatic Plants,* Fassett constructed the keys so they were understandable to the novice botanist and used illustrations extensively to aid the investigator.

At the University of Wisconsin–Madison, Dr. Fassett taught courses on aquatic plants, grasses, advanced taxonomy, and Flora of Wisconsin and its Conservation. He was instrumental in establishing the university arboretum, and he worked with Leopold on the restoration of this land. Fassett personally collected over 28,000 specimens for the university's herbarium, and as curator he expanded the collection fourfold to include over 380,000 specimens. He was also active in preserving areas of diverse flora in Wisconsin, and he led the effort to set aside the first of the scientific areas—the first state-sponsored natural area protection program in the country. These areas are now managed by the Wisconsin Department of Natural Resources under the State Natural Areas (SNA) Program. Currently, the SNA Program administers about 700 sites encompassing more than 360,000 acres of land and water. SNAs protect outstanding examples of native natural communities, significant geological formations, and archaeological sites.

A Manual of Aquatic Plants was extended and revised in 2000 as *Aquatic and Wetland Plants of Northeastern North America,* volumes 1 and 2 by Garrett E. Crow and C. Barre Hellquist. This book covers the more than one thousand species of native and naturalized aquatic and wetland plants found in eastern North America, and it includes keys, geographical ranges, and preferred habitats of these plants.

Frederick and Frances Hamerstrom (begins with personal account by Radomski): I had occasion to visit the Hamerstroms in the 1970s and 1980s. As kids growing up near the confluence of the Wisconsin and Little Plover Rivers, my brothers, Dale Gawlik (a boy from up the road), and I would explore the natural world. Dale Gawlik's grandmother was an avid birder, and she would collect and freeze the large numbers of chipmunks we killed during our bowhunting excursions in the oak/hazelnut woodlots of Portage County, one of Wisconsin's sand counties. I remember visiting the Hamerstroms at this time; I was a passenger, along with frozen chipmunks, being driven across the flat Buena Vista Marsh and catching glimpses of the drainage ditches I fished with my grandfather. As a kid, I thought the Hamerstroms' life a dream (and it still seems that way). Of course they were old back then (in their sixties), and even though I was just a boy, I thought of them as an eccentric and magical couple. They had a great horned owl on the front porch, a large, old home filled with animal specimens, no running water but a cast-iron hand pump on the back porch, and a prairie falcon in their barn. They would go out birding and inspire people to become involved with prairie chickens by getting them to witness these birds on their booming (breeding) grounds. As a college student, I was lucky to visit with them again. On one occasion, while accompanying Dale Gawlik, who worked for the Hamerstroms conducting mice and raptor surveys on his way to become an ecologist, we stopped by as they were leaving for a family event in Boston. They were dressed up in formal eastern wear (they noted their unusual dress, stressing the necessity to please family members), yet spent time to speak with us about our studies and experiences. I had little knowledge and understanding of their past, and I wondered how one could grow up to be like them.

Frances (1907–1998) and Frederick Hamerstrom (1909–1990) were highly regarded ecologists known for their work on prairie chickens of Wisconsin's sand counties. Both grew up in Boston, Fran as the only child of a wealthy family. They met each other as college students. (Fran attended Smith College and Frederick, Dartmouth College. Later, Frederick would graduate from Harvard.) In 1931, the two were married in secret while traveling in Florida. (A formal wedding was held in Boston several months later.) They both shared a love of nature and sought an education in wildlife. From 1932 to 1935 they attended Iowa State University under Paul Errington, a colleague of Aldo Leopold. Frederick received his master's degree and Fran her bachelor's degree studying pheasant nesting, winter ecology of bobwhite quail, and great horned owl diets. After graduating they went north to Necedah, Wisconsin, where Frederick was employed as a game manager for the U.S. government under the Central Wisconsin Game Project. They studied sandhill cranes, prairie chickens, and whitetail deer winter habits, including movements, population size, diets, and food supplies. The Hamerstroms were also wildlife research fellows at the University of Wisconsin under Aldo Leopold. In 1940 Fran earned her master's degree—the only woman to earn a degree under Leopold. In 1941 Frederick earned his doctorate under Leopold. After graduating, the Hamerstroms moved to Michigan. Frederick was the curator of the Edwin S. George Reserve of the University of Michigan, located near Ann Arbor. The reserve is used for research and education in the natural sciences and consists of 1,300 acres of oak-hickory forests and grasslands. During World War II, Frederick served as an aviation physiologist and teacher of survival techniques in the U.S. Army Air Corps and Fran worked as a medical technician in Beaumont Hospital.

The Hamerstroms returned to Wisconsin in 1949 when they accepted positions to lead the newly formed Prairie Grouse Management Research Unit for the State of Wisconsin, headquartered in Plainfield. For the next two decades they would conduct research on prairie chickens and other birds on the marshes and grasslands of the sand counties. Frederick published sixty-nine technical and scientific articles and was an editor for the Wildlife Society, the Raptor Research Foundation, and many of Fran's publications. Fran published over one hundred professional articles and ten books, notably *Harrier, Hawk of the Marshes* published in 1986 by the Smithsonian Institution Press. The Hamerstroms developed plans for acquiring and managing scattered areas of grass- and brushlands critical for a self-sustaining population of prairie chickens. They led efforts to preserve and protect the habitat of the marshland birds they studied by organizing citizens to raise money to purchase or lease lands exceeding thirteen thousand acres. After retiring in 1972, the two continued their work on Buena Vista marshland birds as adjunct professors for the University of Wisconsin–Stevens Point. They exposed hundreds of people, from ornithology students to amateur bird watchers, to the wonders of the rare and magnificent birds of the Buena Vista Marsh. Their legacy can be enjoyed today by exploring the marshes and grasslands of central Wisconsin.

For additional biographical information about the Hamerstroms read the biography written by Corneli (2002).

Charles R. Van Hise: It is interesting that Aldo Leopold rarely mentioned Dr. Van Hise's work. Charles Van Hise (1857–1918) was a leading geologist and president of the University of Wisconsin–Madison from 1903 to 1918. He was born just south of Madison, Wisconsin, and he was the first person to receive a PhD from the University of Wisconsin–Madison. After graduating, he remained to become a professor of mineralogy and geology. Dr. Van Hise also worked for the U.S. Geological Survey from 1909 to 1918 and published extensively on geology, including detailing the geological (and iron ore) formations of the Lake Superior region (Minnesota, Ontario, Michigan, and Wisconsin). As president of the University, Van Hise expanded the university system and courses offered, and he established its medical college. He was a well-known advocate of conservation—conservation in the sense of eliminating or reducing reckless waste of natural resources and not in the sense of preservation or protection of wilderness or natural areas. Van Hise had a deep concern about soil loss, as did Leopold.

In 1910, Dr. Van Hise wrote *The Conservation of Natural Resources in the United States.* It became an often-cited treatise on conservation issues at the beginning of the twentieth century; the book systematically discussed the state of different natural resources from water to energy sources. Van Hise stressed the importance of conserving metals and fuels, as he recognized that the county was growing quickly and many held the false belief that these resources were inexhaustible. He quantified the limits of gas, oil, and coal, and he was concerned that once used they were gone forever. It is safe to say that his recommendations to save these resources for "succeeding generations through unending years to come" have been dismissed as unreasonable and too farsighted. He recommended that coal be used only when absolute necessary, and he encouraged reducing waste and finding substitutes. William Stanley Jevons, an economist, deduced in 1865 (*The Coal Question*) that increases in coal efficiency (i.e., less waste) would only increase coal use due to reduced costs and increasing demand. This has become known as the Jevons paradox. Recently, economists have added economic depth to this phenomenon by exploring both micro- and macroeconomic levels leading to the Khazzoom-Brookes postulate that states that improvements in energy efficiency at the individual level (micro), which are rational and justified, lead to higher energy consumption at the community level (macro). While Van Hise may have been unaware of economic forces leading to increases in coal use, he was aware of other fields of science related to coal. Van Hise cited Svante Arrhenius, a Swedish scientist who in 1896 hypothesized that increases in carbon dioxide from burning of coal and other fuels would increase global temperatures (i.e., the greenhouse effect). Arrhenius's 1896 calculations predicted that a twofold increase in carbon dioxide would mean a 5–6C temperature increase. (The 2007 estimate from the Intergovernmental Panel on Climate Change indicated that a doubling of carbon dioxide in the atmosphere would likely mean a 2–4C increase.)

In his treatise, Van Hise had a mixed record on conservation facts. He was correct in predicting a severe shortage of oil, but wrong in the prediction of the depletion of natural gas reserves within a few years after 1910. Van Hise was a proponent of hydropower and spoke of the benefits of the Mississippi River reservoirs in Minnesota, the Wisconsin River reservoirs in Wisconsin, and the power potential of the Niagara River (where additional hydropower stations would later be built). He recognized that the headwaters of these reservoirs had to be maintained in forest to preserve the streams feeding them. Van Hise was incorrect in his belief that many rivers could be improved with channeling; the field of hydrology has documented the large negative consequences of this activity. It is fair to say that he was correct in stating that the conservation of soil is the greatest of all the problems. He logically stated, "the problem is to make each acre take care of its own rainfall" using vegetation cover, contour plowing, use of terraces, and retention of forest on steep slopes and along streams. (Aldo Leopold would later also state the need for conservation practice on each and every parcel of land.) However, Van Hise failed to see the benefits of wetlands to the hydrologic cycle and overestimated the value of their drainage. He was for vaccination, against child labor, for the eight-hour workday, and an advocate of eugenics (which was popular in the early twentieth century). Finally, Van Hise's statements about high energy density fuels (e.g., oil, natural gas, and coal) only being available for "a small fraction of the time we look forward to as the life of this nation" remain undetermined.

Like Leopold, Van Hise believed that conservation was, in large part, the responsibility of the individual. He stated that our duty was to leave the natural resources to "our descendants as nearly intact as possible." His view of a patriot was certainly different than many who strive for more materials and independence today. "The period in which individualism was patriotism in this country has passed by; and the time has come when individualism must become subordinate to responsibility to the many," said Van Hise. He believed in social responsibility and "to be fair to this generation and considerate of generations to come." Indeed, this is wisdom for the ages.

Balance of Nature: Leopold noted the shortcoming of this mental model of how ecosystems function. The term is still broadly used. Most ecologists now make no illusions to any balance in nature; rather their mental model has expanded to include resilience, multiple-equilibrium conditions or natural states, abrupt shifts in abundance and species composition, and unpredictable systems (Holling 1973; Berkes and Ross 2013).

Leopold's Summary Precept: From *A Sand County Almanac:* "A thing is right when it tends to preserve the integrity, stability, and beauty of the biotic community. It is wrong when it tends otherwise." J. Baird Callicott noted that the "when" should be interpreted as "if," not as "if and only if," as Leopold himself spoke of "accretions" and a "step in a sequence." Leopold wished us to expand our ethics. This general rule means that we should assess the merits of an action to include potential consequences to any human and the community of other animals, plants, water, and the land.

CHAPTER 4. SIGURD OLSON AND PROTECTING WILDERNESS

Sigurd F. Olson's Life: The biographical information about Olson was based on the biography written by Backes (1997).

Calvin Rutstrum: A lover of wilderness and man of many skills, he became a serious author after writing a camping manual for the Lake Hubert Camps in Minnesota, which evolved into *The New Way of the Wilderness* (1958).

Voyageurs National Park: This national park was created with the hard work of environmental leaders in Minnesota, including Sigurd Olson, Minnesota governor Elmer Lee Anderson, Udert W. Hella, Ed Chapman, Robert Watson, Wayne Judy, Martin Kellogg, Wallace Dayton, Wheelock Whitney, Jack Evertt, Dave Zentner, and many others.

Ian McHarg: McHarg (1969) taught urban planners and land-use specialists a systematic, geographic approach to designing communities that fostered the protection of valuable environmental assets and amenities. His work had a profound impact on landscape architecture all over the world. With the aid of geographical information systems (GIS), his approach is now widely used by natural resource managers and government planners. His approach was of ecological sensitivity and guiding development towards taking the path of least ecological resistance.

CHAPTER 5. WILLIAM WHYTE AND HUMAN HABITAT

William H. Whyte's Life: Biographical information about Whyte was gathered from his book *A Time of War,* from *The Essential William H. Whyte* (edited by LaFarge 2000), Birch (1986), and *The Humane Metropolis* (edited by Platt 2006). Whyte's cited Fortune articles can be found in LaFarge (2000). It was Birch (1986) who first called Whyte the Observation Man, and it was Birch who wrote: "Whyte is more than a writer. He is an activist, persuasively marketing his recommendations, translating words into laws and laws into livable environments. Few states in the nation have not been influenced by his conservation work; few downtowns of our largest cities have been left untouched by his findings on the use of public space."

Laurance Spelman Rockefeller: Laurance S. Rockefeller was an important conservationist in the late twentieth century. To gain a deeper appreciation of his life and work, read *Laurance S. Rockefeller: Catalyst for Conservation* (1997) by Robin W. Winks.

Cluster Development and Conservation Subdivisions: A conventional subdivision is a form of subdivision where lots are spread evenly throughout a parcel with little regard for natural features or areas. This is compared to cluster development and conservation subdivisions where lots are clustered and natural areas are protected. Our perspective to rural lakeshore planning and design is in line with conservation design principles, adapted and refined for lakeshore living. Holly Whyte was a strong advocate for returning to this form of rural residential subdivision as the primary design approach for new rural developments. Randall Arendt, landscape planner, has picked up the cause of advocating for conservation subdivisions in the United States and Canada, and he has designed numerous conservation subdivisions that have been environmentally and economically successful. He has written several books and articles that guide developers, governments, and communities on this better form of subdivision (Arendt 1996a, 1996b, 1999, 2003, and 2010). This method of development is characterized by clustering homes adjacent to permanently preserved natural areas. With regard to open space, conservation subdivisions can be compared to golf course developments. However, for conservation subdivisions, community recreational areas and critical natural areas are identified and protected. These protected areas account for over half the subdivision. Then buildable areas are identified and a majority of the lots and homes are clustered around these protected areas. Generally, conservation subdivision designs attempt to create greater community value through natural area or open-space amenities for homeowners and protection of natural resources, while allowing for the residential densities consistent with prevailing densities. Site designs incorporate standards of low-impact development and protection of trees, shoreline buffers, nearshore areas, unique resources, and scenic vistas. These developments use stormwater designs that emphasize on-site retention and infiltration through the preservation of naturally vegetated riparian areas, use of pervious surfaces, rain gardens, and swales.

Conservation subdivisions are an important tool used to provide better lots for homeowners while protecting water quality and wildlife, promoting economic development, and creating open space for recreational use (Arendt 1996a; Wenger and Fowler 2001). Conservation subdivisions also have additional benefits. They may create a greater sense of community, and they allow more interaction with the outdoor environment. People often find these developments more attractive than conventional subdivisions. Open spaces provide walking and biking trails, play areas, and community gathering places. Protected natural areas mean lower development costs, preservation of wildlife habitat, and less pollution runoff into lakes and wetlands. This often leads to both higher property values and higher community value, which strengthens local economies. There are now hundreds of conservation subdivisions across North America, and the paradigm is changing to one in which new rural lakeshore subdivisions are designed with regard to natural features and extensive use of common areas and open space. We should look back to the past and return to the fine tradition of designing neighborhoods with natural features protected and enjoyed by all.

Value of Natural Areas and Conservation Subdivisions: People have long preserved certain natural areas because they provide public value or because they possess rare features. For lakeshores, natural areas are vital for the survival of many species of native flora and fauna. Not only do natural areas provide social amenities for the community such as places to play or relax or socialize, they have economic benefits. Preserved natural areas that are risky for development can reduce a community's susceptibility to floods and fires. Natural areas or open space can also enhance the surrounding land values. Numerous studies have demonstrated an economic value of natural area preservation to individuals and the community (Beaton 1988; National Park Service 1993; Moscovitch 2007; Sander and Polasky 2009; Trust for Public Land 2009).

Within conservation subdivisions, developers can often build at full residential densities, and they often sell lots at a premium because many of us prefer living next to permanently protected natural areas (Nassauer et al. 2004). Development costs may be lower for conservation subdivisions due to much less grading of the site and to shorter and narrower streets than are found in conventional subdivisions, and more compact layouts may also result in shorter sewer and water connections and arterial roads (Caraco et al. 1998). Public service costs of compact conservation-designed developments may also be lower than the cost for large-lot developments (Center for Watershed Protection 1996; Center for Watershed Protection 1998). Conservation subdivisions can also reduce long-term maintenance costs. For example, open space maintained in a natural condition costs up to five times less to maintain than lawns (Schueler 2000).

Several studies have shown that homes in conservation subdivisions appreciate in value more rapidly than homes in conventional subdivisions. Lacy (1990) compared a conventional subdivision with a conservation subdivision. After twenty years, Lacy found that the conservation subdivision lots had appreciated to values 13 percent higher than the conventional subdivision lots that had much larger lot sizes. The price differential was likely due to the recreational amenities and protected natural areas in the conservation subdivision.

Conservation subdivisions can be a valuable tool for protecting water quality and wildlife habitat (Milder and Clark 2011). In a study comparing conservation and conventional subdivisions, streams downstream of conservation subdivisions had lower concentrations of total suspended solids, phosphorus, and nitrate than those downstream of conventionally developed catchments (Nassauer et al. 2004). Caraco et al. (1998) estimated that phosphorus runoff and export may be reduced 60 percent and nitrogen export may be reduced 45 percent using conservation subdivisions and better site designs over conventional developments. The use of natural area wetlands and forests for water filtration can be substantially less expensive than stormwater management based on pipes and ponds or human-made wetlands. Vulnerable natural features can be incorporated within the development, instead of within a private lot, as with conventional subdivisions. Protecting natural areas upfront ensures that these essential community assets continue to provide ecological services. Berke et al. (2003) evaluated fifty matched pairs of conservation and conventional developments across the United States and found that the new urban development practices (e.g., conservation subdivisions and communities designed with low-impact development practices) were more likely to protect and restore sensitive areas, restore degraded stream environments, and provide a more compact alternative to sprawl than conventional developments. Milder et al. (2008) found that conservation subdivisions reduced negative impacts of land development compared to conventional subdivisions. Large conservation subdivisions have the potential to protect important wildlife habitat in the shoreland (Milder 2007). Finally, if planned with neighboring conservation developments and consistent with a comprehensive plan, such developments can preserve wildlife corridors that facilitate animal movement (Arendt 1996a).

CHAPTER 6. ASSET PRESERVATION

Structure Setbacks from Lakes, Rivers, and Wetlands: The proximity principle is key to understanding the need for structure setbacks. Structure setbacks, in association with shoreline buffers, are important because (1) they enhance a site's ability to absorb water before it is conveyed to public waters, (2) they are a basic safeguard for wastewater pollution, (3) they help preserve natural shorelines and aesthetic qualities, and (4) they prevent unsafe building close to the water to minimize damage from fluctuating water levels and bank/bluff erosion. Effectiveness of a structure setback for removal of runoff pollution is a function of distance and slope. As setback decreases, pollution removal decreases. Large setbacks are generally necessary to provide long-term water-quality protection.

CHAPTER 7. ASSET CREATION

Landscape Reconstruction: Thompson and Sorvig (2008) provide a thorough guide to ecologically sensitive landscape reconstruction.

Shoreline Restoration: *Lakescaping for Wildlife and Water Quality* by Henderson et al. (1998) is an outstanding reference guide for restoring shorelines. A lawn to lake, with only the shallow roots of turf holding the soil, leads to bank erosion; whereas the deeper roots of native vegetation reduce bank erosion and enhance aeration and water infiltration. The book describes how to plant and what species of native plants to reestablish along lakes and streams. It also outlines good stewardship practices for lakeshore homeowners.

New Urbanism: For those interested in city planning and design, after reading William Whyte's *City: Rediscovering the Center* (1988) and Jane Jacobs's *The Death and Life of Great American Cities* (1961), we suggest the following:

- *The Smart Growth Manual* by Duany et al. (2009).
- *Suburban Nation: The Rise of Sprawl and the Decline of the American Dream* by Duany et al. (2001).
- *The Option of Urbanism: Investing in a New American Dream* by Leinberger (2008).
- *The Architecture of Community* by Krier (2009).
- *Seven Rules for Sustainable Communities: Design Strategies for the Post-Carbon World* by Condon (2010).
- *Resilient Cities: Responding to Peak Oil and Climate Change* by Newman et al. (2009).

CHAPTER 8. CONNECTING PEOPLE AND THINGS

Streets and Roads: Stefanovic (2012) outlined the factors that one should consider when a community needs to build a road. ASCE (2001) and AASHTO (1994) have recommended residential streets as narrow as twenty-two feet in width. Swift et al. (2006) noted that narrow residential streets have been shown to be the safest, as they slow traffic and reduce vehicular crashes. Interestingly, Dutzik et al. (2011) determined that gasoline taxes do not pay the full cost of highway construction and maintenance. Streets are the biggest contributors of impervious cover in residential developments, and Stone (2004) recommended reduced street widths and street networks that more closely follow a grid pattern (with a five hundred feet block size) to reduce impervious cover in residential areas.

As we think about streets, we need to think about roads. Charles Marohn, an engineer and co-founder and President of Strong Towns, has written extensively on roads and streets (Marohn 2012), and he stated "roads move people between places while streets provide a framework for capturing value within a place." Unfortunately as Marohn has demonstrated in many North American places, we tend to build more roads than streets. Around our lakes we need streets—streets that are narrow, are designed for people movement not vehicles or rapid transit, and include places to stop, sit, and enjoy the views and open space.

Road and Highway Removal: Using case studies, Cervero (2009) summarized the benefits of converting elevated freeways to boulevards or public open space. The Congress for the New Urbanism identified the top twelve highways for conversion to boulevards (CNU 2012).

CHAPTER 9. CULTURE AND GOVERNANCE

A Placed Person: The conservationist writer Paul Gruchow (1947–2004) spoke of taking a large group of Minnesota high school students to a nearby lake. He asked them to identify the plants along the shore. Gruchow (1995) reported: "The majority of the students could name no more than two or three. The dandelion was the only plant they all knew. They didn't recognize cattails. Most of them couldn't tell the difference between a willow tree and a cottonwood tree. They have wandered and played along that lakeshore for a lifetime, utterly blind to it."

Extinction of Experience: Additional readings on this topic and on the need for conserving natural areas in urban landscapes include Miller and Hobbs (2002), Pyle (2003), and Miller (2005, 2006, and 2008).

Sense of Place: Time may also be critical in the development of a sense of place. "Place can acquire deep meaning for the adult through the steady accretion of sentiment over the years," wrote geographer Yi-Fu Tuan in *Space and Place* (1977).

Science of Sense of Lake Place: Social studies on the sense of place as it relates to lakeshore living include Jorgensen and Stedman (2001, 2006), Stedman (2003, 2006), Stedman and Hammer (2006), Stedman et al. (2007), and Schroeder (2009). Schroeder surveyed Minnesota residents on their attachments to lakes. She found that lakeshore residents with high lake recreational activity had higher place attachment and expressed higher commitment for lake protection than those lakeshore residents that did not engage in regular lake activities.

Community: Wendell Berry, poet and essayist, discussed the merits of a viable community and the shortcomings of the unrestrained free-market economy in an essay titled "The Idea of a Local Economy" (2001) and in a Jefferson Lecture (2012).

Planning for Growth: Marohn (2012) notes that the suburban development pattern may be a land-use Ponzi scheme. He articulates a compelling message that the continued failure to understand the long-term liabilities of the infrastructure required for suburbs jeopardizes the vibrancy of our towns.

Urban Growth Boundary: Reduced amenities beyond the boundary often resulted in lower land and housing values compared to those within the boundary (Cho et al. 2006; Jun 2006). The effectiveness of this urban containment strategy is poorly understood, and the results appear mixed (Bengston et al. 2004; Jun 2004; Gennaio et al. 2009; Hepinstall-Cymerman et al. 2013).

Edward M. Bassett: Edward Bassett (1863–1948) was a lawyer, and he also served one term in the U.S. House of Representatives (1903–1905). As a citizen, Bassett involved himself in various appointed roles in government, especially in New York City. He served on school boards and the New York Public Service Commission. His work on zoning began in earnest when he served as chair of the New York City Heights of Building Commission and chairman of the Zoning Commission, where in 1916 he developed the first zoning ordinance in the country. In the 1910s and 1920s, he was active in the annual National Conference on City Planning (Bassett 1917, 1923). At the Eleventh National Conference on City Planning in 1919, Bassett spoke to the legal standing of zoning when he stated:

> We have tried to establish zoning on the firmest possible foundation of safety. We long ago concluded that it could not progress if brought about by condemnation and payment. We considered it must be done under the police power or not at all. The police power can be invoked only for safety, health, morals, and general convenience of the community. The courts cling pretty closely to the first three reasons. There must be a clear relation between the specific zoning requirement and the health, safety and morals of the community, or else the court will say that the attempted use of power is arbitrary and amounts to taking property for a public use without just compensation.

At the Fourteenth National Conference on City Planning in 1922, Bassett, in response to a question, clarified the limits of zoning as he viewed at the time by stating:

> Zoning the land and buildings of a city is not taking them for a public use. It is placing protection upon land and buildings for the common benefit. Zoning in the United States is done under the police power. No money payment is made to the property owners. He has a right, however, to be protected against confiscation or arbitrary or unequal treatment. He has a right to insist that the zoning shall be reasonable and for the benefits of the community. The comprehensive plan must have a relation to health, safety, morals and the general welfare. It cannot be confiscatory or for aesthetic purposes or merely to enhance the value of land or buildings.

In 1922 Bassett was appointed by Herbert Hoover to the Advisory Committee on Zoning to the U.S. Department of Commerce.

Edward Bassett is considered the father of American zoning due to his early work in New York City and his dedication to informing city officials across the county on the merits of comprehensive zoning. Bassett was also very interested in transportation within cities. He remembered New York City when it had no public transportation except horse cars, and he was a strong supporter of rapid transit. Bassett said, "rapid transit lines are the skeleton of the city." Interestingly, while he advocated that cities plan for rapid transit, his views on automobile transportation turned out to be harmful to city centers. In a highly regarded published article, he provided the name and vision for a new type of highway:

> We are more and more feeling the need of a new kind of thoroughfare—one which will be like a highway for both pleasure and business vehicles, but which will be like a parkway in preventing the cluttering-up of its edges. We have no name for such a thoroughfare. No law in this country provides for such a novel trafficway. If a name could be given to this new sort of thoroughfare, it would immediately enter into the practice as well as the terminology of city planning. I suggest the name of freeway. This word is short and good Anglo-Saxon. It connotes freedom from grade intersections and from private entrance ways, stores and factories. It will have no sidewalks and will be free from pedestrians. In general, it will allow a free flow of vehicular traffic. It can be adapted to the intensive parts of great cities for the uninterrupted passage of vast numbers of vehicles. (Bassett 1930)

Citizen Guide to Land Use: Hunnsicker and West (2007) provide a good introduction on the land-use decision-making process.

Lakeshore Zoning Economics: Economists at the University of Wisconsin have extensively studied the consequences of zoning on lakeshore property values in northern Wisconsin, and their information is broadly applicable. Zoning has been used to optimize property value, and the first of these economic studies evaluated whether an increase in a minimum lake frontage requirement would increase land value (Spalatro and Provencher 2001). In Wisconsin, the state requires at least one hundred feet of lake frontage per lot in unsewered areas. Several towns in Vilas County (and a few in Oneida County) adopted a two-hundred-foot minimum frontage restriction in their shoreland ordinances, which obviously results in fewer lake homes per lake. Vilas County, in northern Wisconsin, has 1,320 lakes, and water covers about 15 percent of the surface area in the county. The study was based on property sales data collected on 893 undeveloped lakeshore properties from towns in these counties with one-hundred-foot and two-hundred-foot frontage requirements from 1986 to 1995. The economists developed a multivariable regression model to predict the property value per unit of frontage. The analysis indicated that there was neither a price jump for properties with two hundred feet of frontage in towns using the state minimum standard nor a price fall at two hundred feet where the two-hundred-foot standard was required. The two-hundred-foot frontage requirement did appear to have a negative effect on properties with frontage between two hundred and four hundred feet and a positive price jump for four-hundred-foot lots. Most importantly, the two-hundred-foot frontage requirement generally increased the price per foot of undeveloped lake frontage. The model predicted the average expected price under the two-hundred-foot standard to be 21.5 percent greater than the average observed sale price under the one-hundred-foot standard. This research indicated that the zoning requirement, by preserving clean water and natural beauty, generated an economic gain that more than offset the economic loss resulting from the constraints on development. People are willing to pay more to live on a lake that is protected from degradation and lightly developed.

Several studies evaluated the economic consequences of the 1999 lake classification system adopted by Vilas County that grouped lakes into classes based on level of development and ecological sensitivity (Papenfus and Provencher 2006; Lewis et al. 2009). This classification approach followed the example from Minnesota, which in 1970 classified lakes into three classes that then determined the standards for development. Vilas County created development standards that were most protective of lakes with low development and high ecological sensitivity, and they enacted less restrictive development requirements for lakes with already high development and low sensitivity. The primary differences in development standards for the different lake classes were minimum lot size (60,000 square feet for lakes with the most protective regulations and 30,000 square feet for less restrictive lakes) and minimum lakeshore frontage (300 feet for the most protective and 150 feet for the least restrictive). One study modeled the property value related to the minimum frontage requirements to determine if the classification produced negative or positive economic results (Papenfus and Provencher 2006). The data for the study included market sales from 1997 to 2001(before and after the adoption of the lake classification) for both developed and undeveloped parcels. The effect of the classification with the requirement of larger frontages for low development and high sensitivity lakes was a 10 percent reduction in the number of potential lots. The economists concluded that the lake classification and its associated development restrictions increased lakeshore property values, even for the undeveloped parcels that could no longer be subdivided due to a larger lot frontage requirement; the average price effect for these parcels was positive, and 70 percent were estimated to be more valuable with the change. This study also supported the benefit of protective lakeshore zoning requirements: by preserving clean water and natural beauty, the local government generated an economic gain that more than offset the economic loss resulting from the constraints on development.

Another study surveyed thousands of Vilas County lakeshore property owners on their willingness to pay to prevent additional lakeshore development and used simulation models to determine if people buying lakeshore property will sort themselves by environmental conditions and lakeshore frontage prices (Lewis and Provencher 2006). This study found that people had indeed sorted across Vilas County lakes. Survey respondents with large parcels on less developed lakes were more likely to indicate a preference for a reduction in the development on their lakes, and most interesting, this preference was strongest for the recent lakeshore purchasers. The simulations suggested that the lake classification zoning approach might foster people sorting themselves based on different preferences of environmental condition and price (lakes with high environmental amenities having high prices versus lakes with low environmental amenities having lower prices). Without the lake classification and associated zoning controls, the lakes in the county were predicted to become more homogeneous—lakes with lower environmental condition and a reduced price differential between lakes—as higher environmental amenity lakes would develop faster than low amenity lakes.

The last set of economic studies investigated the effect of Vilas County's zoning controls (minimum lake frontage requirements) and the amount of public conservation lands abutting a lake in determining whether a landowner subdivided their property and the ecological effects of such subdivision (Butsic et al. 2010; Lewis 2010; and Butsic et al. 2012). An econometric model was created using a data set of subdividable lakeshore lots from 140 Vilas County lakes that were followed from 1974 through 1998 (a total sample size of 1,575 lots). Researchers found that development consisted of many small subdivisions made by individual property owners, and about 11 percent of parcels were subdivided more than once. While large parcels (those greater than 1,500 feet of frontage) generated about half of all new lakeshore lots, 82 percent of the subdivision plats were due to small developments that generated less than six new lakeshore lots each. The probability of subdivision was higher on large lakes. From 1974 to 1998, lakeshore residential density increased by 24 percent. Interestingly, 85 percent of the subdivisions produced lakeshore lots that exceeded the minimum lake frontage requirement (i.e., new lakeshore lot density was generally lower than that allowed by zoning). The researchers found no evidence that the minimum lake frontage requirements affected the probability of subdivision. However, researchers found evidence that the amount of public land abutting the lake influenced the large parcel property owner's decision on subdividing their property. This result was consistent with the earlier mentioned survey in which some lakeshore property owners indicated a preference for a reduction of development on their lake. Parcel size and policy on public lands abutting lakes were key in determining lakeshore development, such that an increase in public conservation land was predicted to reduce the probability that large parcels would be subdivided. The econometric model that predicted probability of subdivision was then integrated with ecological models to determine if zoning changes could reduce loss of aquatic habitat, improve bluegill growth rates, and reduce the loss of green frog populations with lakeshore development. Researchers found that increasing the minimum frontage zoning standards did reduce the loss of nearshore downed tree aquatic habitat and the loss of green frog populations. However, simulation results of different zoning provisions (minimum lake frontages of one hundred, two hundred, three hundred, and four hundred feet) suggested diminishing returns, with the greatest ecological benefits coming from increasing the minimum lake frontage from one hundred feet to two hundred feet. In addition, the researchers noted that ecological effects from increasing the minimum lake frontage requirement were more likely to occur on lakes that were initially less developed, and they recommended that more protective zoning standards be applied only to these lakes in an effort to maximize environmental benefits. Lastly, researchers found that zoning and land acquisition generally had a small but positive effect on lakeshore property values, and both these policies could be more effective in increasing property value when targeted to specific lakes in the region.

Lakeshore Economics: Shoreline property owners, local governments, and taxpayers benefit economically as a result of the amenities that good lakeshore and shoreline management preserves: clean water, fish and wildlife, and natural beauty (Dempsey 2006). Good water quality is often critical to the tax base and economic assets of an area. Water clarity is strongly related to the price people are willing to pay for lakefront property. In a five-year study of nine hundred shorefront properties on thirty-four lakes in Maine, declining water clarity was shown to reduce lakefront property values and could increase the tax burden of offshore properties (Michael et al. 1996). A three-foot difference in average minimum water clarity was associated with property value declines of up to 22 percent. In a lake-rich township in Maine, it was predicted that a three-foot decline in average minimum water clarity would cause a loss of 5 percent in total property value and likely an equivalent loss in taxes paid (Maine Department of Environmental Protection 1996). A similar study showed a direct relationship between property values and water clarity for Minnesota lakes (Krysel et al. 2003). Lakes with clearer water were associated with higher property values, while lakes with less clarity were associated with lower property values. The study looked at 1,205 properties sold on 37 lakes in the communities of Aitkin, Brainerd, Grand Rapids, Walker, Park Rapids, and Bemidji. This study found that a three-foot increase in water clarity had an economic worth of $50 per foot for lake frontage, or about $5,000 for a typical property with one hundred feet of lakeshore. And a three-foot decrease in clarity led to a much higher proportionate loss in economic worth, averaging $70 per foot of lake frontage. This study and those in Maine are evidence that protecting water quality of lakes is important in maintaining the economic assets of a region.

Scenic quality attributes are generally higher for lightly developed than poorly or overly developed landscapes, and especially for lakes. Stedman and Hammer (2006) found that when people perceived the lake around which they owned property as more developed, they were more likely to see that lake as polluted. The effect of shoreline development on the perception of polluted water was as strong as that of the actual measure of water greenness (i.e., chlorophyll or total phosphorus concentration). Therefore, the type of development and the perception of the development are both important if an area is to continue to attract tourist and resort business. Shoreline buffers can increase property values (Wenger and Fowler 2000). Buffers protect water quality, which has immense economic value. Clean public waters are also valuable for recreation and tourism and are important factors in attracting new businesses and residents.

Dziuk and Heiskary (2003) estimated that ten lakes in Itasca County, Minnesota, generated an estimated total income of $7 million in direct consumer purchases and gross output of wages and other expenses, whereas real estate taxes for shoreland properties paid to the county amounted to about $333,000. They concluded that the substantial income from lakes is adequate justification for keeping them in a healthy state through use of best management practices. They hoped that recognition of the amount of income from healthy lakes would lead to a greater commitment on the part of local officials in supporting efforts to take better care of such resources.

Joseph L. Sax: Joseph L. Sax is currently an emeritus professor of law at the University of California, Berkeley, and a fellow of the American Academy of Arts and Sciences. He has taught law at the University of Colorado, University of Michigan, Stanford University, University of Utah, and University of Paris. He wrote the seminal work on the public trust doctrine applied to natural resource protection law (Sax 1970). His recent writings on land-use regulation include Sax (1989, 1996, 2005, 2010, and 2011).

Other Lake Regulations: Lakeshore residents should be aware of numerous regulations other than zoning law. Many governments regulate or prohibit the destruction of aquatic plants growing in lakes and rivers to balance the needs of riparian property owners to obtain reasonable access to public waters and protection of aquatic plant communities that provide critical habitat for fish and wildlife. Several governments regulate the size and placement of docks and piers. State and provincial governments also commonly regulate activities that alter the lake bottom, such as dredging, adding fill, breakwalls, retaining walls, riprap, and the placement of structures, such as boathouses.

While retaining walls and rock riprap can reduce erosion, these shoreline alterations can also reduce habitat for many species (Jennings et al. 1999; Trial et al. 2001; Meadows et al. 2005; Gabriel and Bodensteiner 2012). Many lakeshores and even some wetlands have been ringed with rock, often where it is not needed (Wehrly et al. 2012). Rock riprap protection should be limited to the amount necessary to protect eroding areas from wave action, and it should only be used where an established erosion problem cannot be controlled through the use of vegetation and slope stabilization.

CHAPTER 10. SYSTEM CHANGING

Surface Water Management: Booth et al. (2004) found no evidence that the impacts of urban development can be fully alleviated. They recommended several reasonable actions to rehabilitate water resources:

- cluster development to protect most of the natural vegetative cover, especially in headwater areas and around streams and wetlands, so that riparian buffers remain intact;
- limit watershed imperviousness, either through minimal development or by reducing the "effective" impervious area through the widespread infiltration of stormwater;
- mimic natural flow frequencies and durations, not just control peak discharges;
- protect riparian buffers and wetland zones, and minimize road and utility crossings;
- begin landowner stewardship programs that recognize the unique role of adjacent private property owners in rehabilitating, maintaining, or degrading lake and stream health.

Runoff volume reduction is becoming a standard in stormwater management (Hirschman et al. 2008; Schueler 2008). Numerous studies have demonstrated that low-impact development (LID) design and extensive use of infiltration and bioretention basins can effectively reduce runoff volume and pollution transport (U.S. EPA 2000; Cheng et al. 2004; Selbig and Bannerman 2008).

The effectiveness of rain gardens has been confirmed by many studies (e.g., Selbig and Balster 2010). The rain gardens and bioretention systems that line some of the streets in Maplewood, Minnesota, have been shown to clean and infiltrate runoff, replenish underground aquifers, and increase property value and beauty (Nassauer et al. 2001). Dietz and Clausen (2005) found that rain gardens treated 99 percent of the toxins in runoff captured. A network of rain gardens in a Burnsville, Minnesota, neighborhood reduced stormwater runoff by over 90 percent during a 0.75 inch rainfall (Barr Engineering 2004). Monitoring data from 2004 showed that the rain gardens achieved an 80 percent reduction in runoff volume in forty-nine rain events. Most basins drained dry within three to four hours. Today there are many good references written for homeowners on rain garden construction and use of other rainwater harvesting and infiltration techniques.

The National Research Council (2009) comprehensively reviewed stormwater management within cities in the United States. The Council determined that urban areas generally have failed to address stormwater runoff volumes. The result of not addressing stormwater volumes at individual project sites has resulted in large downstream problems, from blow-outs of stream banks to large volumes of pollutants discharged into urban lakes. The report suggests that municipalities focus on reducing impervious surface cover, manage stormwater by watersheds, and appropriately regulate future land development. The U.S. EPA (2010) summarized the benefits of using green infrastructure for stormwater management at the watershed, neighborhood, and site level scales. Hager et al. (2013) studied the challenges and benefits of retrofitting stormwater best management practices within a dense neighborhood in Baltimore, Maryland. They found improvements in water quality and increases in recreational use, but they noted the difficulty in planning, implemented and maintaining stormwater retrofits in older, dense cities.

Machu Picchu: Research about water management on Machu Picchu includes Wright et al. (2000) and Wright (2004).

Need for Retooling our Water Use and Wastewater Management: See Glennon (2009) for a discussion on the evidence of a water crisis in North America. Leaking pipes and illicit sewage discharges can be a significant pollution source in many communities. In two urban watersheds Lilly et al. (2012) estimated that illicit discharges were the primary phosphorus loading source. Anger et al. (2013) quantified triclosan and its toxic derivatives in Minnesota lake sediments. A good scientific introduction to endocrine disruptors has been written by Colborn et al. (1997). Brodin et al. (2013) found that a common prescription drug to treat anxiety, which is found in rivers below wastewater treatment plants, altered perch behavior and their feeding rates. The impacts of wastewater discharge on Canadian lakes and rivers were summarized by Holeton et al. (2011).

The U.S. EPA (2011 in association with the Combined Heat and Power Partnership; 2012) has documented the energy potential from using anaerobic digesters at wastewater treatment facilities. The anaerobic digester technology is not new, and it is currently used within many wastewater systems (e.g., human, dairy farms, and pig farms). Facilities need to link advanced digesters with equipment to generate electricity. This approach is now economical and wide adoption would reduce air and water pollution. A history of urban farming was written by Lawson (2005), and the use of city places for advanced food production was discussed by Nordahl (2009). In the Minneapolis-Saint Paul metropolitan area, Fissore et al. (2011) found phosphorus inputs dominated by human diet, detergents, and pet food, and Baker (2011) estimated that the deliberate export of phosphorus for agricultural use may support about half of the food production required in the Twin Cities watershed.

Watershed-Based Governance: Ruhl et al. (2003) discussed the concept of a comprehensive state-regional-local institutional structure. A review of watershed planning efforts across the United States was conducted by Lehman et al. (2012). They found that some collaborative efforts have made progress in managing watersheds, but attainment of water-quality standards or goals has been rare.

Civilizations and Failures of Governance: The collapse of a government prior to the modern era often resulted in the collapse of the society or even the civilization. Today governments collapse regularly, but international organizations generally step in to assist the development of a new government. While this appears to be rather far afield, it is instructive to learn from those who have studied failures. Scientists Jared Diamond and Joseph Tainter studied societies where governance has failed dramatically. Diamond (2005) discussed the fates of Easter Islanders, Maya society, the Norse in Greenland, and many others. He concluded that a society needs to remain resilient, which requires extensive efforts to minimize environmental damage. Tainter (1990 and additional research) came to understand that complexity is the result of problem solving. As a society places greater demands on resources, which generates more problems, it adapts by creating new jobs, institutions, bureaucracies, and specializations. Tainter found that collapse occurs as a result of declining returns on investment into further social complexity. This persistent increase in complexity is nearly impossible to escape; societies become stuck in this positive feedback loop. The solutions to escape the feedback loop are generally socially unacceptable and temporary (e.g., substantially reducing population and economic growth rates, periodically simplifying social and economic systems by a deliberate reset or reboot). But as the economist Herbert Stein (1916–1999) said, "If something cannot go on forever, it will stop."

Better Zoning: Szold and Carbonell (2002), Leinberger (2008), and Newman et al. (2009) discussed the need to promote compact, mixed-use, transit-oriented development and more efficient land-use patterns due to increasing energy costs. There are many environmental benefits of more efficient land use, including less pollution on a per capita basis (Campoli and MacLean 2007). Elliot (2008) also advocated for flexibility in dealing with nonconforming lots of record. He stated that it is reasonable to recognize that some nonconformity in older, built-out areas may be addressed without the need for a variance. Finally, Elliot stated that there is little to be gained by going through a process that may politicize the decision by the local government, which he termed "late-in-the-game NIMBYism."

Process for Negotiating with Developers: Schoenbauer (2007) outlined the approach and made the case that negotiation can be a viable alternative alongside a traditional prescriptive ordinance track.

Extent of Lawns: See Milesi et al. (2005).

Rethinking Invasive Species Management: Radomski and Goeman (1995) and Rahel (2002a) researched the homogenization of freshwater animals. Sigurd Olson introduced smallmouth bass into Minnesota lakes, and several studies have demonstrated the consequences of smallmouth bass introductions on other fish species (MacRae and Jackson 2001; Vander Zanden et al. 2004; Aiken et al. 2012). Dreissenid mussel, including zebra mussel (*Dreissena polymorpha*) and quagga mussel (*D. rostriformis bugensis*), colonizations have large economic consequences and alter ecosystem food webs (Higgins et al. 2011), but no viable whole-lake control exists.

Besides Rosenzweig's *Win-Win Ecology* (2003), additional thoughts and ideas on invasive species management include the following: Chew and Laubichler (2003), Lodge et al. (2006), Breining (2008), Keulartz and Van der Weele (2008), Davis (2009), Zhang and Boyle (2010), Strayer (2010), Davis et al. (2011), and Powell et al. (2013). Strayer (2010) noted that eutrophication and disturbance favor invasive aquatic plant colonization, and he predicted that the number of lake invasions of alien species would greatly increase, with the rate only slowed by education and regulatory programs. Water transparency often decreases following herbicide treatment of aquatic plants (O'Dell et al. 1995; Valley et al. 2006; Wagner et al. 2007). Schlaepfer et al. (2011) noted that many nonnative species provide important ecological value. Schlaepfer et al. (2012) made a compelling case for a more balanced approach in dealing with nonnative species.

Economic Systems: Water-related payments for ecosystem services systems have been discussed by Smith et al. (2006), Brauman et al. (2007), and Goldman-Benner et al. (2012). Can market forces replace morality and personal responsibility with regard to water pollution? Sandel (2012) raised fairness and corruption concerns about use of markets as a replacement for personal responsibility, but should not businesses pay for their pollution (negative externalities) in circumstances where regulation has proven inadequate or inappropriate (e.g., agricultural runoff)? Is it fair that a corporation sends their runoff pollution downstream while another property owner infiltrates rainwater? Gaylord Nelson, the late Wisconsin governor and U.S. senator, thought we should look at the relation of the economy to the environment differently. He stated, "The economy is a wholly owned subsidiary of the environment. . . . The economy is, after all, just a subset within the ecological system" (Nelson et al. 2002).

Walkability: Speck (2012) documented ten elements necessary to create a walkable city, starting with designing streets for people first rather than for increasing car traffic.

Conservation Prioritization Tools: As threats to lakes continue to mount, it is becoming increasingly important to identify and conserve high-priority areas. Two common systematic approaches for conservation prioritization are system-based models and value-based models. One of the major strengths of system-based models is that they require us to think deeply about a system by writing down our mental models of how we believe the system functions. However, we often do not have system models that can accurately identify where in the lakeshed specific good management practices should be applied or that have the ability to simulate alternative land management actions and predict consequences at specific locations in the watershed. Value-based models use a compilation of individual criteria of valuable landscape features and aggregated criteria with an objective function to prioritize places within the landscape for conservation. Although there are some shortcomings of using value models over system models (value models only allow exploration of tradeoffs and optimization, and they do not provide guidance on what practices should be implemented where), the use of value models is an efficient method for prioritizing places for protection or restoration (Moilanen et al. 2009). Value models can help achieve a multiple benefits goal by identifying areas that optimize benefits by accounting for what the community values. The use of an additive benefits objective function in the value models allows the retention of high-quality occurrences of as many conservation features as possible while reducing interference between competing land uses. Value models also can be used in a public participation process, whereby participants can decide on what natural resource features are valued and the ranking of those valued features. In addition, value models are simple concepts that are easy to explain and apply at the local government scale. Benedict and McMahon (2006) outline ten principles of good systematic conservation design.

Development Impact Fees: Nelson et al. (2008) reviewed these one-time fees for new development and presented cases where they have been used successfully.

Using Social Psychology to Advance Social Norms: Often people change their behavior when they are repeatedly confronted with facts, and most people are receptive to the subtle messaging of social norms. The late psychologist B. F. Skinner (1904–1990), in *Upon Further Reflection* (1987), noted that we often need contrived reinforcements for good behavior. Progress has been on such reinforcements. Thaler and Senstein (2008) used psychology research to provide guidance on how governments, corporations, and organizations can nudge or manipulate people to make better choices. They provided examples of choice settings that influenced outcomes. For example, changing the organ default option to automatically opting in instead of out, has resulted in high participation in an organ donor program. The psychologist Daniel Kahneman, in *Thinking, Fast and Slow* (2011) explained how we think. He described our modes of cognitive function and our cognitive biases. His insights on these topics, along with his development of prospect theory, provide understanding in how people chose between probabilistic alternatives, gains and losses, and thinking fast and thinking slow.

Everett Rogers (1931–2004), in *Diffusion of Innovations* (2003), studied how new ideas and practices permeate society and why some innovations fail to catch on. Rogers noted that when an innovation successfully spreads out through a community it generally follows an S-shaped pattern of adoption over time. An innovation is initially slow to catch on, then adoption speeds up as word spreads, and finally there is a leveling off of adoption as the innovation saturates the population. The rate of adoption is usually a function of the innovation. If the innovation has advantages over existing practices, is compatible with values and experiences, is less complex, is easy to experiment with, and has high visibility to others in the community, then it is more likely to have a high rate of adoption. Rogers identified several important steps in the diffusion of innovation, and he found that to succeed in adoption a person needs

(1) to learn about an innovation, (2) to be persuaded about the merits of the innovation, (3) to try out the innovation often with experimentation, and (4) confirmation of the innovation's merits from peers (positive reinforcement). Rogers also grouped people into adopter categories: innovators, early adopters, early majority, late majority, and laggards. Naturally, one first needs to focus on the innovators and early adopters to advance the S-shaped adoption curve. Open-minded community opinion leaders and professionals that promote innovations help to speed up diffusion. Rogers's review of studies on successful and failed adoptions provides the reader, the change agent, with some great insights to how to increase the probability of making good things happen.

Marketing science will also help us in our cause. A cause has a communication strategy and a systematic method to bring out the best in people. Several psychologists have developed effective frameworks. Two popular psychologists and their specific frameworks include Robert Cialdini on influence and persuasion and Doug McKenzie-Mohr on fostering sustainable behavior. Cialdini (2003, 2006) identified several critical factors that can be used to bring out the best in people; they include the use of give-and-take, getting oral or written commitment, providing social proof of the action, persuasion from someone well-liked or an authority figure, and communicating the facts of rarity or scarcity of natural resources to generate interest and demand for action. McKenzie-Mohr (2011) also uses some of these critical factors in a market approach he teaches. He notes that information is key and that we need to go beyond just providing it. Often there are barriers, both physical and mental, that inhibit us from altering our behavior to one that is more sustainable. McKenzie-Mohr found that it is first necessary to identify barriers. For example, Wisconsin researchers surveyed lakeshore property owners on their barriers to restoring lakeshore habitat. First, owners were concerned about obstructed lake views and reduced ability to see young children playing down by the lake. Second, owners were concerned that restoration of a shoreline buffer would preclude a sandy beach. Third, people were concerned about insects and other invertebrate pests within the natural vegetation of the buffer.

Once barriers are identified, McKenzie-Mohr's framework proceeds to the development of a strategy to change behavior. The strategy may include the use of oral or written commitments, the use of prompts to remind and encourage people to continue an activity, making the social norm visible to others, communicating the information persuasively (in part by using Cialdini's frameworks), use of incentives, and removal of barriers. In the example of restoring lakeshore habitat, the removal of barriers could include several strategies. First, we can provide information on where, what, and how to plant a shoreline buffer that provides views down to the lake and frames the view so that owners can still see children and family playing down by the lake. Second, we can confirm and assure lakeshore residents that beach areas and access corridors can be included with a restoration design they choose. Third, we can provide advice on the use of mulched paths and lawn-edged corridors through shoreline buffer areas in their design to minimize pest encounters. This information, as well as the benefits of shoreline buffers (protecting water quality, providing fish and wildlife habitat, protecting shoreline from erosion, keeping geese off lawns, etc.), would be easily communicated to neighbors. A valuable additional step in McKenzie-Mohr's framework includes monitoring and evaluating your strategy. For restoration of lakeshore habitat, the Minnesota DNR has developed a simple tool called *Score Your Shore* (2010) that lakeshore residents or lake associations can use to evaluate either a personal or a community restoration. There are other science-based marketing programs available, most of which include collaborative learning (e.g., you learn about barriers, your neighbor learns about benefits), developing good relationships within a community, diffusion of innovation, and evaluation. Examples of successful programs can be found in McKenzie-Mohr et al. (2011).

Our conservation problems are social problems that are first and foremost about challenges in changing human behavior. Rittel and Webber (1973) declared that these kinds of problems are inherently "wicked" problems to solve, and they caution scientists, engineers, and planners to be wary of the intoxicating effects of attempting to apply the modern-classical model of governmental planning, which includes extensive reliance on forecasts and simulations. They assert that use of such methods is bound to fail, as these tools were developed to deal with "tame" problems, not public policy problems. Collaborative approaches to address conservation may be the most effective way to begin working toward changing human behavior and perceptions. Brown (2010) stated, "since wicked problems are generated by the society in which they are set, their resolution will necessarily involve changes in the society that produced them." Partial and complete solutions to specific wicked problems are found and implemented every day. Sometimes, after taking a fresh look at the problem with an open mind, a simple but unintuitive solution becomes clear. Meadows (2008) provided an example related to information flows: "Change comes first from stepping outside the limited information that can be seen from any single place in the system and getting an overview. From a wider perspective, information flows, goals, incentives, and disincentives can be restructured so that separate, bounded, rational actions do add up to results that everyone desires. It's amazing how quickly and easily behavior changes can come, with even slight enlargement of bounded rationality, by providing better, more complete, timelier information."

CHAPTER 11. OUR LAKE, OUR RESPONSIBILITY

Community Design (theory and philosophy): For additional reading see Van Assche and Verchraegen (2008), Van Assche et al. (2012), and Van Assche et al. (2013).

References

AASHTO (American Association of State Highway and Transportation Officials). 1994. A policy on geometric design of highways and streets. AASHTO, Washington, DC.

Aiken, J. K., S. C. Findlay, and F. Chapleau. 2012. Long-term assessment of the effect of introduced predatory fish on minnow diversity in a regional protected area. Canadian Journal of Fisheries and Aquatic Sciences 69:1798–1805.

Albert, M. R., G. Chen, G. K. MacDonald, J. C. Vermaire, E. M. Bennett, and I. Gregory-Eaves. 2010. Phosphorus and land-use changes are significant drivers of cladoceran community composition and diversity: an analysis over spatial and temporal scales. Canadian Journal of Fisheries and Aquatic Sciences 67:1262–1273.

American Farmland Trust. 2010. Assess the cost of community services. American Farmland Trust, Washington, DC.

Anderson, D. R. 2007. Model based inference in the life sciences: a primer on evidence. Springer, New York.

Anderson, K. A., T. J. Kelly, R. M. Sushak, C. A. Hagley, D. A. Jensen, and G. M. Kreag. 1999. Summary report with tables of survey responses on public perceptions of the impacts, use, and future of Minnesota lakes: results of the 1998 Minnesota lakes survey. University of Minnesota Sea Grant and Minnesota Department of Natural Resources, St. Paul.

Anger, C. T., C. Sueper, D. J. Blumentritt, K. McNeill, D. R. Engstrom, and W. A. Arnold. 2013. Quantification of triclosan, chlorinated triclosan derivatives, and their dioxin photoproducts in lacustrine sediment cores. Environmental Science and Technology (in press).

Anton, P. A. 2005. The economic value of open space: implications for land use decisions. Wilder Research, St. Paul, MN.

Arendt, R. G. 1996a. Conservation design for subdivisions: a practical guide to creating open space networks. Island Press, Washington, DC.

———. 1996b. Creating open space networks. Environment & Development, American Planning Association. May/June.

———. 1999. Growing greener: putting conservation into local plans and ordinances. Island Press, Washington, DC.

———. 2003. Linked landscapes: creating greenway corridors through conservation subdivision design strategies in the northeastern and central United States. Landscape and Urban Planning 68:241–269.

———. 2010. Envisioning better communities: seeing more options, making wiser choices. American Planning Association's Planners Press and the Urban Land Institute, Washington, DC.

Arrow, K., B. Bolin, R. Costanza, P. Dasgupta, C. Folke, C. S. Holling, B.-O. Jansson, S. Levin, K.-G. Maler, C. Perrings, and D. Pimentel. 1995. Economic growth, carrying capacity, and the environment. Science 268:520–521.

ASCE (American Society of Civil Engineers). 2001. Residential streets, 3rd edition. Edited by W. M. Kulash, in association with National Association of Home Builders and Urban Land Institute. Urban Land Institute, Washington, DC.

Asplund, T. 1997. Investigations of motor boat impacts on Wisconsin's lakes. Wisconsin Department of Natural Resources, Publication PUB-SS-927, Madison.

———. 2000. The effects of motorized watercraft on aquatic ecosystems. Wisconsin Department of Natural Resources, Publication PUB-SS-948-00, Madison.

Asplund, T. R., and C. M. Cook. 1997. Effects of motor boats on submerged aquatic macro-phytes. Journal of Lake Reservoir Management 13:1–12.

———. 1999. Can no-wake zones effectively protect littoral zone habitat from boating disturbance? LakeLine 19(1):16–18.

Backes, D. 1997. A wilderness within: the life of Sigurd Olson. University of Minnesota Press, Minneapolis.

Baker, L. A. 2011. Can urban P conservation help prevent the brown devolution? Chemosphere 84(6):779–784.

Baker, L. A., J. E. Schussler, and S. A. Snyder. 2008a. Drivers of change for lakewater clarity. Lake and Reservoir Management 24:30–40.

Baker, L. A., B. Wilson, D. Fulton, and B. Horgan. 2008b. Disproportionality as a framework to target pollution reduction from urban landscapes. Cities and the Environment 1(2):article 7.

Barr Engineering. 2004. Burnsville rainwater gardens. Land and Water Conservation (Sept./Oct.):47–51.

Barten, J. 2005. Stormwater management for lakeshore and near lakeshore homeowners. LakeLine 25(1):21–24.

Bassett, E. M. 1917. Constitutional limitations on city planning powers. Proceedings of the ninth national conference on city planning, Kansas City, MO, May 7–9:199–227.

———. 1923. Present attitude of courts toward zoning. Proceedings of the fifteenth national conference on city planning, Baltimore, MD, April 30, May 1–2:115–144.

———. 1930. The freeway: a new kind of thoroughfare. The American City (February):95.

Beachler, M. M., and D. F. Hill. 2003. Stirring up trouble? Resuspension of bottom sediments by recreational watercraft. Lake and Reservoir Management 19:15–25.

Beaton, W. P. 1988. The cost of government regulations, volume 2. A baseline study for the Chesapeake Bay critical area. Chesapeake Bay Critical Area Commission, Annapolis, MD.

Becker, G. C. 1983. Fishes of Wisconsin. University of Wisconsin Press, Madison.

Belanger, L., and J. Bedard. 1990. Energetic cost of man-induced disturbance to staging snow geese. Journal of Wildlife Management 54:36–41.

Benedict, M. A., and E. T. McMahon. 2006. Green infrastructure: linking landscapes and communities. Island Press, Washington, DC.

Bengston, D. N., J. O. Fletcher, and K. C. Nelson. 2004. Public policies for managing urban growth and protecting open space: policy instruments and lessons learned in the United States. Landscape and Urban Planning 69:271–286.

Benson, E. R., J. M. O'Neil, and W. C. Dennison. 2008. Using the aquatic macrophyte *Vallisneria americana* (wild celery) as a nutrient bioindicator. Hydrobiologia 596:187–196.

Bentrup, G. 2008. Conservation buffers: design guidelines for buffers, corridors, and greenways. Gen. Tech. Rep. SRS-109, U.S. Department of Agriculture, Forest Service, Southern Research Station, Asheville, NC.

Berke, P. R., J. MacDonald, N. White, M. Holmes, D. Line, K. Oury, and R. Ryznar. 2003. Greening development to protect watersheds: does new urbanism make a difference? Journal of the American Planning Association 69(4):397–413.

Berkes, F., and H. Ross. 2013. Community resilience: toward an integrated approach. Society and Natural Resources 26:5–20.

Bernthal, T. W. 1997. Effectiveness of shoreland zoning standards to meet statutory objectives: a literature review with policy implications. Wisconsin Department of Natural Resources, Publication PUBL-WT-505-97, Madison.

Bernthal, T. W., and S. A. Jones. 1997. Shoreland management program assessment. Wisconsin Department of Natural Resources, Publication PUBL-WT-506-97 and PUBL-WT-507-97, Madison, WI.

Berry, W. 2001. The idea of a local economy. Orion Magazine 20(1):28–37.

———. 2012. It all turns on affection. Jefferson Lecture, National Endowment for the Humanities, Washington, DC.

Birch, E. L. 1986. The observation man. Planning (March):4–8.

Blann, K. L., J. L. Anderson, G. R. Sands, and B. Vondracek. 2009. Effects of agricultural drainage on aquatic ecosystems: a review. Critical Reviews in Environmental Science and Technology 39:909–1001.

Booth, D. B., J. R. Karr, S. Schauman, C. P. Konrad, S. A. Morley, M. G. Larson, and S. J. Burges. 2004. Reviving urban streams: land use, hydrology, biology, and human behavior. Journal of the American Water Resources Association 40(5):1351–1364.

Boyer, D. R. 1965. Ecology of basking habit in turtles. Ecology 46:99–118.

Brabec, E., S. Schulte, and P. L. Richards. 2002. Impervious surfaces and water quality: a review of current literature and its implications for watershed planning. Journal of Planning Literature 16(4):499–514.

Brauman, K. A., G. C. Daily, T. K. Duarte, and H. A. Mooney. 2007. The nature and value of ecosystem services: an overview highlighting hydrological services. Annual Review of Environment and Resources 32:67–98.

Breining, G. 2008. Alien invasion. Minnesota Monthly 42(6). Available: http://www.minnesota monthly.com/media/Minnesota-Monthly/June-2008/Alien-Invasion/ (August 2013).

Brodin, T., J. Fick, M. Jonsson, and J. Klaminder. 2013. Dilute concentrations of a psychiatric drug alter behavior of fish from natural populations. Science 339(6121):814–815.

Brown, D. E., and N. B. Carmony, editors. 1995. Aldo Leopold's southwest. University of New Mexico Press, Albuquerque (originally published in 1990 as Aldo Leopold's wilderness, Stackpole Books, Harrisburg, PA).

Brown, D. G., K. M. Johnson, T. R. Loveland, and D. M. Theobald. 2005. Rural land-use trends in the conterminous United States, 1950–2000. Ecological Applications 15:1851–1863.

Brown, V. A. 2010. Collective inquiry and its wicked problems. Pages 61–81 in V.A. Brown, J.A. Harris, and J.Y. Russell, editors. Tackling wicked problems through the transdisciplinary imagination. Earthscan, Washington, DC.

Bryan, M. D., and D. L. Scarnecchia. 1992. Species richness, composition, and abundance of fish larvae and juveniles inhabiting natural and developed shorelines of a glacial Iowa lake. Environmental Biology of Fishes 35:329–341.

Buell, L. 2003. Writing for an endangered world: literature, culture, and environment in the U.S. and beyond. Belknap Press of Harvard University Press, Cambridge, MA.

Burnham, K. P., and D. R. Anderson. 2002. Model selection and multi-model inference. Springer, New York.

Butsic, V., J. W. Gaeta, and V. Radeloff. 2012. The ability of zoning and land acquisition to increase property values and maintain largemouth bass growth rates in an amenity rich region. Landscape and Urban Planning 107:69–78.

Butsic, V., D. J. Lewis, and V. C. Radeloff. 2010. Lakeshore zoning has heterogeneous ecological effects: an application of a coupled economic-ecological model. Ecological Applications 20:867–879.

Callicott, J. B. 1987. Companion to *A Sand County Almanac:* interpretive and critical essays. University of Wisconsin Press, Madison.

Callicott, J. B, and E. T. Freyfogle, editors. 1999. Aldo Leopold: for the health of the land. Island Press/Shearwater Books, Washington DC.

Campoli, J., and A. S. MacLean. 2007. Visualizing density. Lincoln Institute of Land Policy, Cambridge, MA.

Cappiella, K., and T. Schueler. 2001. Crafting a lake protection ordinance. Pages 751–768 in Urban lake management. Center for Watershed Protection, Ellicott City, MD.

Caraco, D., R. Claytor, and J. Zielinski. 1998. Nutrient loading from conventional and innovative site development. Center for Watershed Protection, Silver Spring, MD.

Caron, J. A., and W. L. Robinson. 1994. Responses of breeding common loons to human activity in upper Michigan. Hydrobiologia 279/280:431–438.

Carpenter, S. R. 2005. Eutrophication of aquatic ecosystems: bistability and soil phosphorus. Proceedings of the National Academy of Sciences of the United States of America 102:10002–10005.

———. 2008. Phosphorus control is critical to mitigating eutrophication. Proceedings of the National Academy of Sciences of the United States of America 105:11039–11040.

Carpenter, S. R., N. F. Caraco, D. L. Correll, R. W. Howarth, A. N. Sharpley, and V. H. Smith. 1998. Nonpoint pollution of surface waters with phosphorus and nitrogen. Ecological Applications 8:559–568.

Carpenter, S. R., and R. C. Lathrop. 2008. Probabilistic estimate of a threshold for eutrophication. Ecosystems 11:601–613.

Carpenter, S. R., and twenty coauthors. 2007. Understanding regional change: comparison of two lake districts. BioScience 57:323–335.

Carson, R. 1962. Silent spring. Reissued 2002 (40th anniversary edition), Houghton Mifflin, New York.

Castelle, A. J., C. Conolly, M. Emers, E. D. Metz, S. Meyer, M. Witter, S. Mauerman, T. Erickson, and S. Cooke. 1992. Wetland buffers: use and effectiveness. Washington State Department of Ecology, Olympia, WA.

Castelle, A. J., A. W. Johnson, and C. Conolly. 1994. Wetland and stream buffer size requirements—a review. Journal of Environmental Quality 23:878–882.

Center for Watershed Protection. 1996. Site planning for urban stream protection. Metropolitan Washington Council of Governments, Ellicott City, MD.

———. 1998. Better site design: a handbook for changing development rules in your community. Center for Watershed Protection, Ellicott City, MD.

Cervero, R. 2009. Transportation infrastructure and global competitiveness: balancing mobility and livability. Annuals of the American Academy of Political and Social Science 626(1):210–225.

Cheng, M.-S., L. S. Coffman, Y. Zhang, and Z. J. Licsko. 2004. Comparison of hydrological responses from low impact development with conventional development. Pages 419–430 in M. Clar, D. Carpenter, J. Gracie, and L. Slate, editors. Protection and restoration of streams: proceedings of the symposium, June 23–25, 2003, Philadelphia, PA. American Society of Civil Engineers, Reston, VA.

Chew, M. K., and M. D. Laubichler. 2003. Natural enemies: metaphor or misconception? Science 301:52–53.

Chi, G., and D. W. Marcouiller. 2012. Recreational homes and migration to remote amenity-rich areas. Journal of Regional Analysis and Policy 42(1):47–60.

Cho, S.-H., Z. Chen, S. T. Yen, and D. B. Eastwood. 2006. Estimating effects of an urban growth boundary on land development. Journal of Agricultural and Applied Economics 38(2):287–298.

Christensen, D. L., B. R. Herwig, D. E. Schindler, and S. R. Carpenter. 1996. Impacts of lakeshore residential development on coarse woody debris in north temperate lakes. Ecological Applications 6:1143–1149.

Cialdini, R. B. 2003. Crafting normative messages to protect the environment. Current Directions in Psychological Science 12(4):105–109.

———. 2006. Influence: the psychology of persuasion, revised edition. Harper Collins, New York.

Clady, M. D. 1976. Change in abundance of inshore fishes in Oneida Lake, 1916–1970. New York Fish and Game Journal 23:73–81.

Clair, T. A., I. F. Dennis, and R.Vet. 2011. Water chemistry and dissolved organic carbon trends in lakes from Canada's Atlantic provinces: no recovery from acidification measured after 25 years of lake monitoring. Canadian Journal of Fisheries and Aquatic Sciences 68:663–674.

Clark, T., and D. Euler. 1984. Vegetation disturbance caused by cottage development in central Ontario. Journal of Environmental Management 18:229–239.

CNU (Congress for the New Urbanism). 2012. Freeways without futures. Congress for the New Urbanism, Chicago. Available: http://www.cnu.org/sites/www.cnu.org/files/final_2012_freeways_without_futures_3.pdf (August 2013).

Cohen, P., and J. Stinchfield. 1984. Shoreland development trends, Minnesota Department of Natural Resources, Shoreland Update Project Report 4, St. Paul.

Colborn, T., D. Dumanoski, and J. P. Myers. 1997. Our stolen future: are we threatening our fertility, intelligence, and survival?—a scientific detective story. Plume, New York.

Commoner, B. 1971. The closing circle: nature, man, and technology. Alfred A. Knopf, New York.

Condon, P. M. 2010. Seven rules for sustainable communities: design strategies for the post-carbon world. Island Press, Washington, DC.

Corneli, H. M. 2002. Mice in the freezer, owls on the porch: the lives of naturalists Frederick and Frances Hamerstrom. University of Wisconsin Press, Madison.

Correl, M., J. Lillydahl, and L. Singell 1978. The effect of greenbelts on residential property values: some findings on the political economy of open space. Land Economics 54(2): 207–17.

Crompton, J. L. 2007. The role of the proximate principle in the emergence of urban parks in the United Kingdom and in the United States. Leisure Studies 26:213–234.

Crow, G. E., and C. B. Hellquist. 2000. Aquatic and wetland plants of northeastern North America, volumes 1 and 2. University of Wisconsin Press, Madison.

Cuffney, T. F., R. A. Brightbill, J. T. May, and I. R. Waite. 2010. Responses of benthic macroinvertebrates to environmental changes associated with urbanization in nine metropolitan areas. Ecological Applications 20:1384–1401.

Daily, G. C., S. Alexander, P. R. Ehrilich, L. Goulder, J. Lubchenco, P. A. Matson, H. A. Mooney, S. Postel, S. H. Schneider, D. Tilman, and G. M. Woodwell. 1997. Ecosystem services: benefits supplied to human societies by natural ecosystems. Issues in Ecology 2:1–16.

Daily, G. C., P. R. Ehrlich, and M. Alberti. 1996. Managing Earth's life support systems: the game, the players, and getting everyone to play. Ecological Applications 6:19–21.

Dale, V. H., S. Brown, R. A. Haeuber, N. T. Hobbs, N. Huntly, R. J. Naiman, W. E. Riebsame, M. G. Turner, and T. J. Valone. 2000. Ecological principles and guidelines for managing the use of land. Ecological Applications 10:639–670.

Davis, M. A. 2009. Invasion biology. Oxford University Press, New York.

Davis, M. A., M. K. Chew, R. J. Hobbs, A. E. Lugo, J. J. Ewel, G. J. Vermeiji, J. H. Brown, M. L. Rosenzweig, M. R. Gardener, S. P. Carroll, K. Thompson, S. T. A. Pickett, J. C. Stromberg, P. Del

Tredici, K. N. Suding, J. G. Ehrenfeld, J. P. Grime, J. Mascaro, and J. C. Briggs. 2011. Don't judge species by their origins. Nature 474:153–154.

DeLuca, W. V., C. E. Studds, L. L. Rockwood, and P. P. Marra. 2004. Influence of land use on the integrity of marsh bird communities of Chesapeake Bay, USA. Wetlands 24:837–847.

Dempsey, D. 2006. Minnesota calling: conservation facts, trends and challenges. Minnesota Campaign for Conservation, St. Paul.

Dennis, J. 1986. Phosphorus export from a low-density residential watershed and an adjacent forest watershed. Lake and Reservoir Management 2:401–407.

Desbonnet, A., V. Lee, P. Pogue, D. Reis, J. Boyd, J. Y. Willis, and M. Imperial. 1995. Development of coastal vegetated buffer programs. Coastal Management 23:91–109.

Diamond, J. 1999. Guns, germs and steel: fates of human societies. W.W. Norton, New York.

———. 2005. Collapse: how societies choose to fail or succeed. Viking, New York.

Dickson, T., and J. R. Tomelleri. 2008. The great Minnesota fish book. University of Minnesota Press, Minneapolis.

Diebel, M. W., J. T. Maxted, D. M. Robertson, S. Han, and M. J. Vander Zanden. 2009. Landscape planning for agricultural nonpoint source pollution reduction III: assessing phosphorus and sediment reduction potential. Environmental Management 43:69–83.

Dietz, M. E., and J. C. Clausen. 2005. Saturation to improve pollution retention in a rain garden. Environmental Science and Technology 40:1335–1340.

Dillon, P. J., W. A. Schneider, R. A. Reid, and D. S. Jeffries. 1995. Lakeshore capacity study: Part 1—test of effects of shoreline development on the trophic status of lakes. Lake and Reservoir Management 8:121–129.

Downing, J. A., Y. T. Prairie, J. J. Cole, C. M. Duarte, L. J. Tranvik, R. G. Striegl, W. H. McDowell, P. Kortelainen, N. F. Caraco, J. M. Melack, and J. Middelburg. 2006. The global abundance and size distribution of lakes, ponds, and impoundments. Limnology and Oceanography 51:2388–2397.

Duany, A., E. Plater-Zyberk, and J. Speck. 2001. Suburban nation: the rise of spawl and the decline of the American dream. North Point Press, New York.

Duany, A., J. Speck, and M. Lydon. 2009. The smart growth manual. McGraw-Hill, New York.

DuBois, R. 2005. Damselflies of the north woods. Kollath-Stensaas Publishing, Duluth, MN.

Dutzik, T., B. Davis, and P. Baxandall. 2011. Do roads pay for themselves? Setting the record straight on transportation funding. U.S. PIRG Education Fund, Boston, MA.

Dziuk, H., and S. Heiskary. 2003. Local economic impact of healthy lakes. LakeLine 23(3):21–23.

Ehrenfeld, J. 2008. Sustainability by design: a subversive strategy for transforming our consumer culture. Yale University Press, New Haven, CT.

Elias, J. E., and M. W. Meyer. 2003. Comparisons of undeveloped and developed shorelands, northern Wisconsin, and recommendations for restoration. Wetlands 23:800–816.

Elliot, D. L. 2008. A better way to zone: ten principles to create more livable cities. Island Press, Washington, DC.

Engel, S., and J. L. Pederson Jr. 1998. The construction, aesthetics, and effects of lakeshore development: a literature review. Wisconsin Department of Natural Resources, Research Report 177, Madison, WI.

Evers, D. C., L. J. Savoy, S. R. DeSorbo, D. E. Yates, W. Hanson, K. M. Taylor, L. S. Siegel, J. H. Cooley Jr., M. S. Bank, A. Major, K. Munney, B. F. Mower, H. S. Vogel, N. Schoch, M. Pokras, M. W. Goodale, and J. Fair. 2008. Adverse effects from environmental mercury loads on breeding common loons. Ecotoxicology 17:69–81.

Evers, D. C., J. D. Paruk, J. W. McIntyre and J. F. Barr. 2010. Common Loon (*Gavia immer*), The Birds of North America Online (A. Poole, Ed.). Ithaca: Cornell Lab of Ornithology; Available: http://bna.birds.cornell.edu/bna/species/313 (August 2013). .

Fang, X., P. C. Jacobson, H. G. Stefan, S. R. Alam, and D. L. Pereira. 2012. Identifying cisco refuge lakes in Minnesota under future climate scenarios. Transactions of the American Fisheries Society 141:1608–1621.

Fassett, N. C. 1957. A manual of aquatic plants, 2nd edition. University of Wisconsin Press, Madison.

Ficke, A. D., C. A. Myrick, and L. J. Hansen. 2007. Potential impacts of global climate change on freshwater fisheries. Reviews in Fisheries Biology and Fisheries 17:581–613.

FISRWG (Federal Interagency Stream Restoration Working Group). 1998. Stream corridor restoration: principles, processes, and practices. GPO Item No. 0120-A; SuDocs No. A 57.6/2:EN 3/ PT.653, Washington, DC.

Fissore, C., L. A. Baker, S. E. Hobbie, J. Y. King, J. P. McFadden, K. C. Nelson, and I. Jakobsdottir. 2011. Carbon, nitrogen, and phosphorus fluxes in household ecosystems in the Minneapolis-Saint Paul, Minnesota, urban region. Ecological Applications 21:619–639.

Flader, S., and J. B. Callicott, editors. 1991. The river of the mother of God and other essays by Aldo Leopold. University of Wisconsin Press, Madison.

Foley, J. 2010. Hold the red herrings, please. Momentum Magazine 2(2): Directors Note.

Foley, J. A., C. Monfreda, N. Ramankutty, and D. Zaks. 2007. Our share of the planetary pie. Proceedings of the National Academy of Sciences of the United States of America 104:12585–12586.

Francis, T. B., and D. E. Schindler. 2006. Degradation of littoral habitats by residential development: woody debris in lakes of the Pacific Northwest and Midwest, United States. Ambio 35(6):274–280.

———. 2009. Shoreline urbanization reduces terrestrial insect subsidies to fishes in North American lakes. Oikos 118:1872–1882.

Francis, T. B., D. E. Schindler, J. M. Fox, and E. Seminet-Reneau. 2007. Effects of urbanization on the dynamics of organic sediments in temperate lakes. Ecosystems 10:1057–1068.

Francl, K. E., and G. D. Schnell. 2002. Relationships of human disturbance, bird communities, and plant communities along the land–water interface of a large reservoir. Environmental Monitoring and Assessment 73(1):67–93.

Fraser, J. D., L. D. Frenzel, and J. E. Mathiesen. 1985. The impact of human activities on breeding bald eagles in north-central Minnesota. Journal of Wildlife Management 49:585–592.

Frost, P. C., and A. L. Hicks. 2012. Human shoreline development and the nutrient stoichiometry of aquatic plant communities in Canadian Shield lakes. Canadian Journal of Fisheries and Aquatic Sciences 69:1642–1650.

Gabriel, A. O., and L. R. Bodensteiner. 2012. Impacts of riprap on wetland shorelines, Upper Winnebago pool lakes, Wisconsin. Wetlands 32:105–117.

Garn, H. S. 2002. Effects of lawn fertilizer on nutrient concentration in runoff from lakeshore lawns, Lauderdale Lakes, Wisconsin. U.S. Geological Survey, Water Resources Investigations Report 02-4130.

Garn, H. S., D. L. Olson, T. L. Seidel, and W. J. Rose. 1996. Hydrology and water quality of Lauderdale Lakes, Walworth County, Wisconsin, 1993–94. U. S. Geological Survey, Water Resources Investigations Report 96-4235.

Garn, H. S., D. M. Robertson, W. J. Rose, and D. A. Saad. 2010. Hydrology, water quality, and response to changes in phosphorus loading of Minocqua and Kawaguesaga Lakes, Oneida County, Wisconsin, with special emphasis on effects of urbanization. U.S. Geological Survey, Scientific Investigations Report 2010-5196.

Garrison, P. J., G. D. LaLiberte, and B. P. Ewart. 2010. The importance of water level changes and shoreline development in the eutrophication of a shallow seepage lake. Proceedings of the Academy of Natural Sciences of Philadelphia 160:113–126.

Garrison, P. J., and R. S. Wakeman. 2000. Use of paleolimnology to document the effect of lake shore development on water quality. Journal of Paleolimnology 24:369–393.

Genkai-Kato, M., and S. R. Carpenter. 2005. Eutrophication due to phosphorus recycling in relation to lake morphometry, temperature, and macrophytes. Ecology 86:210–219.

Gennaio, M.-P., A. M. Hersperger, and M. Gurgi. 2009. Containing urban sprawl—evaluating effectiveness of urban growth boundaries set by the Swiss land use plan. Land Use Policy 26(2):224–232.

Gergel, S. E., E. M. Bennett, B. K. Greenfield, S. King, C. A. Overdevest, and B. Stumborg. 2004. A test of the environmental Kuznets curve using long-term watershed inputs. Ecological Applications 14:555–570.

Glennon, R. 2009. Unquenchable: America's water crisis and what to do about it. Island Press, Washington, DC.

Goldman-Benner, R. L., S. Benitez, T. Boucher, A. Calvache, G. Daily, P. Kareiva, T. Kroeger, and A. Ramos. 2012. Water funds and payments for ecosystem services: practice learns from theory and theory can learn from practice. Oryx 46(1):55–63.

Graczky, D. J., and S. R. Greb. 2006. Soil data at sites near Geneva Lake, Lake Geneva, Wisconsin, and Long Lake, near New Auburn, Wisconsin. U.S. Geological Survey, Open-File Report 2006-1191.

Graczky, D. J., R. J. Hunt, S. R. Greb, C. A. Buchwald, and J. T. Krohelski. 2003. Hydrology, nutrient concentrations, and nutrient yields in nearshore areas of four lakes in northern Wisconsin. U.S. Geological Survey, Water Resources Investigations Report 03-4144.

Gratton, C., and M. J. Vander Zanden. 2009. Flux of aquatic insect productivity to land: comparison of lentic and lotic ecosystems. Ecology 90:2689–2699.

Grubaugh, J. W., and R. V. Anderson. 1988. Spatial and temporal availability of floodplain habitat: long-term changes at Pool 19, Mississippi River. American Midland Naturalist 119:402–411.

Gruchow, P. 1995. Grass roots: the universe of home. Milkweed Editions, Minneapolis, MN.

Haberl, H., K.-H. Erb, F. Krausmann, V. Gaube, A. Bondeau, C. Plutzar, S. Gingrich, W. Lucht, and M. Fisher-Kowalski. 2007. Quantifying and mapping the human appropriation of net primary production in earth's terrestrial ecosystems. Proceedings of the National Academy of Sciences of the United States of America 104:12942–12947.

Haberl, H., K.-H. Erb, F. Krausmann, and M. McGinley. 2010. Global human appropriation of net primary production (HANPP). Encyclopedia of Earth. Environmental Information Coalition, National Council for Science and the Environment, Washington, DC. Available: http://www.eoearth.org/view/article/153031/ (August 2013).

Hager, G. W., K. T. Belt, W. Stack, K. Burgess, J. M. Grove, B. Caplan, M. Hardcastle, D. Shelley, S. T. A. Pickett, and P. M. Groffman. 2013. Socioecological revitalization of an urban watershed. Frontiers in Ecology and the Environment 11:28–36.

Haines, A. L., T. T. Kennedy, and D. L. McFarlane. 2011. Parcelization: forest change agent in northern Wisconsin. Journal of Forestry 109:101–108.

Hamerstrom, F. 1986. Harrier, hawk of the marshes: the hawk that is ruled by a mouse. Smithsonian Institution Press, Washington DC.

Hardin, G. 1968. The tragedy of the commons. Science 162:1243–1248.

Hatzenbeler, G. R., J. M. Kampa, M. J. Jennings, and E. E. Emmons. 2004. A comparison of fish and aquatic plant assemblages to assess ecological health of small Wisconsin lakes. Lake and Reservoir Management 20:211–218.

Heimberger, M., D. Euler, and J. Barr. 1983. The impacts of cottage development on common loon reproductive success in central Ontario. Wilson Bulletin 95(3):431–439.

Heiskary, S. A., and E. B. Swain. 2002. Water quality reconstruction from fossil diatoms: applications for trend assessment, model verification, and development of nutrient criteria for lakes in Minnesota, USA. Minnesota PCA, St. Paul.

Henderson, C. L., C. J. Dindorf, and F. J. Rozumalski. 1998. Lakescaping for wildlife and water quality. Minnesota Department of Natural Resources, St. Paul.

Henning, B. M., and A. J. Remsburg. 2008. Lakeshore vegetation effects on avian and anuran populations. American Midland Naturalist 161:123–133.

Hepinstall-Cymerman, J., S. Coe, and L. R. Hutyra. 2013. Urban growth patterns and growth management boundaries in the central Puget Sound, Washington, 1986–2007. Urban Ecosystems (in press).

Heraty, M. 1993. Riparian buffer programs: a guide to developing and implementing a riparian buffer program as an urban stormwater best management practice. Metropolitan Washington Council of Governments. U.S. Environmental Protection Agency, Office of Wetlands, Oceans, and Watersheds, Washington, DC.

Hicks, A. L., and P. C. Frost. 2011. Shifts in aquatic macrophyte abundance and community composition in cottage developed lakes of the Canadian Shield. Aquatic Botany 94:9–16.

Higgins, S. N., M. J. Vander Zanden, L. N. Joppa, and Y. Vadeboncoeur. 2011. The effect of dreissenid invasions on chlorophyll and the chlorophyll : total phosphorus ratio in north-temperate lakes. Canadian Journal of Fisheries and Aquatic Sciences 68:319–329.

Hirschman, D., K. Collins, and T. Schueler. 2008. The runoff reduction method. Technical Memorandum, Center for Watershed Protection, Ellicott City, MD.

Hobbs, W. O., J. M. Ranstack Hobbs, T. LaFrancois, K. D. Zimmer, K. M. Theissen, M. B. Edlund, N. Michelutti, M. G. Bulter, M. A. Hanson, and T. J. Carlson. 2012. A 200-year perspective on alternative stable state theory and lake management from a biomanipulated shallow lake. Ecological Applications 22:1483–1496.

Holeton, C., P. A. Chambers, and L. Grace. 2011. Wastewater release and its impacts on Canadian waters. Canadian Journal of Fisheries and Aquatic Sciences 68:1836–1859.

Holling, C. S. 1973. Resilience and stability of ecological systems. Annual Review of Ecology and Systematics 4:1–23.

Holm, B., and D. Ohman. 2007. Cabins of Minnesota. Minnesota Historical Society Press, St. Paul.

Hunnsicker, P., and P. West. 2007. A citizen's guide to influencing local land-use decision. 1000 Friends of Minnesota, St. Paul, and Minnesota Waters, St. Cloud.

Hunt, R. J., S. R. Greb, and D. J. Graczyk. 2006. Evaluating the effects of nearshore development on Wisconsin lakes. U.S. Geological Survey, Fact Sheet 2006-3033.

Hurd, B. 2001. Stirring the mud: on swamps, bogs and the human imagination. Beacon Press, Boston.

Hurlbert, S. H. 2012. Population camel gets its nose into ecologists' tent: hope is high that the rest will follow. The Social Contract 23(1):68–75.

Jacobs, J. 1961. The death and life of great American cities. Random House, New York (republished 1993, Modern Library, New York).

Jacobson, P. C., H. G. Stefan, and D. L. Pereira. 2010. Coldwater fish oxythermal habitat in Minnesota lakes: influence of total phosphorus, July air temperature, and relative depth. Canadian Journal of Fisheries and Aquatic Sciences 67:2002–2013.

Jakes, P. J., C. Schlichting, and D. H. Anderson. 2003. A framework for profiling a lake's riparian area development potential. Journal of Environmental Management 69:391–400.

Jennings, M. J., M. A. Bozek, G. R. Hazenbeler, E. E. Emmons, and M. D. Staggs. 1999. Cumulative effects of incremental shoreline habitat modifications on fish assemblages in north temperate lakes. North American Journal of Fisheries Management 19:18–27.

Jennings, M. J., E. E. Emmons, G. R. Hatzenbeler, C. Edwards, and M. A. Bozek. 2003. Is littoral habitat affected by residential development and land use in watersheds of Wisconsin lakes? Lake and Reservoir Management 19:272–279.

Jiang, L., X. Fang, H. G. Stefan, P. C. Jacobson, and D. L. Pereira. 2012. Oxythermal habitat parameters and identifying cisco refuge lakes in Minnesota under future climate scenarios using variable benchmark periods. Ecological Modelling 232:14–27.

Jorgensen, B. S., and R. C. Stedman. 2001. Sense of place as an attitude: lakeshore owners attitudes toward their properties. Journal of Environmental Psychology 21:233–248.

———. 2006. A comparative analysis of predictors of sense of place dimensions: attachment to, dependence on, and identification with lakeshore properties. Journal of Environmental Management 79:316–327.

Jun, M.-J. 2004. The effects of Portland's urban growth boundary on urban development patterns and commuting. Urban Studies 41(7):1333–1348.

———. 2006. The effects of Portland's urban growth boundary on housing prices. Journal of the American Planning Association 72(2):239–243.

Kahl, R. 1991. Boating disturbance of canvasbacks during migration at Lake Polygan, Wisconsin. Wildlife Society Bulletin 19:242–248.

Kahneman, D. 2011. Thinking, fast and slow. Farrar, Straus, and Giroux, New York.

Kelly, T., and J. Stinchfield. 1998. Lakeshore development patterns in northeast Minnesota: status and trends. Minnesota Department of Natural Resources, Office of Management and Budget Services, St. Paul.

Kelly, V. R., G. M. Lovett, K. C. Weathers, S. E. G. Findlay, D. L. Strayer, D. J. Burns, and G. E. Likens. 2008. Long-term sodium chloride retention in a rural watershed: legacy effects of road salt on streamwater concentration. Environmental Science and Technology 42:410–415.

Keulartz, J., and C. Van der Weele. 2008. Framing and reframing in invasion biology. Configurations 16(1):93–115.

Kling, G. W., K. Hayhoe, L. B. Johnson, J. J. Magnuson, S. Polasky, S. K. Robinson, B. J. Shuter, M. M. Wander, D. J. Wuebbles, D. R. Zak, R. L. Lindroth, S. C. Moser, and M. L. Wilson. 2003. Confronting climate change in the Great Lakes region: impacts on our communities and ecosystems. Union of Concerned Scientists, Cambridge, MA, and Ecological Society of America, Washington, DC. Available: http://www.ucsusa.org/assets/documents/global_warming/greatlakes _final.pdf (August 2013).

Knapp, A. K., and C. D'Avanzo. 2010. Teaching with principles: toward more effective pedagogy in ecology. Ecosphere 1(6):1–10.

Knapton, R., S. Petrie, and G. Herring. 2000. Human disturbance of diving ducks on Long Point Bay, Lake Erie. Wildlife Society Bulletin 28:923–930.

Korschgen, C., and R. Dahlgren. 1992. Human disturbances of waterfowl: causes, effects, and management. Fish and Wildlife Leaflet 13.2.15. Waterfowl Management Handbook. U.S. Fish and Wildlife Service, Washington, DC.

Kramer, D. B., S. Polasky, A. Starfield, B. Palik, L. Westphal, S. Snyder, P. Jakes, R. Hudson, and E. Gustafson. 2006. A comparison of alternative strategies for cost-effective water quality management in lakes. Environmental Management 38:411–425.

Krier, L. 2009. The architecture of community. Island Press, Washington, DC.

Krysel, C., E. Marsh Boyer, C. Parson, and P. Welle. 2003. Lakeshore property values and water quality: evidence from property sales in the Mississippi Headwaters Region. Mississippi Headwater Board, Backus, MN.

Kuhn, T. S. 1996. The structure of scientific revolutions, 3rd edition. University of Chicago Press, Chicago.

Kurlansky, M. 1997. Cod: a biography of the fish that changed the world. Walker, New York.

Lacy, J. 1990. An examination of market appreciation for clustered housing with permanent open space. Center for Rural Massachusetts, Department of Landscape Architecture and Regional Planning, University of Massachusetts, Amherst.

LaFarge, A., editor. 2000. The essential William H. Whyte. Fordham University Press, New York.

Lambert, D., A. Cattaneo, and R. Carignan. 2008. Periphyton as an early indicator of perturbation in recreational lakes. Canadian Journal of Fisheries and Aquatic Sciences 65:258–265.

Landsman, S. J., V. M. Nguyen, L. F. G. Gutowski, J. Gobin, K. V. Cook, T. R. Binder, N. Lower, R. L. McLaughlin, and S. J. Cooke. 2011. Fish movement and migration studies in the Laurentian Great Lakes: research trends and knowledge gaps. Canadian Journal of Fisheries and Aquatic Sciences 37:365–379.

Lenhart, C. F., H. Peterson, and J. Nieber. 2011. Increased streamflow in agricultural watersheds of the Midwest: implications for management. Watershed Science Bulletin 2(1):25–31.

Lant, C., J. B. Ruhl, and S. Kraft. 2008. The tragedy of ecosystem services. BioScience 58:969–974.

Larson, E. R., J. D. Olden, and N. Usio. 2011. Shoreline urbanization interrupts allochthonous subsidies to a benthic consumer over a gradient of lake size. Biology Letters 7:551–554.

Lathrop, R. 2007. Perspectives on the eutrophication of the Yahara Lakes. Lake and Reservoir Management 23:345–365.

Lawson, L. J. 2005. City bountiful: a century of community gardening in America. University of California Press, Berkeley.

Lehman, S., K. Cappiella, J. Schneider, and L. Woodworth. 2012. Tracking progress of watershed planning: two views. Watershed Science Bulletin 3(2):7–20.

Leinberger, C. B. 2008. The option of urbanism: investing in a new American dream. Island Press, Washington, DC.

Leopold, A. 1925. Wilderness as a form of land use. Journal of Land and Public Utility Economics 1(4):398–404.

———. 1931. Report on a game survey of the north central states. Sporting Arms and Ammunition Manufacturer's Institute, Madison, WI.

———. 1932. Game and wild life conservation. Condor 34:103–106.

———. 1933a. The conservation ethic. Journal of Forestry 31:634–643.

———. 1933b. Game management. Charles Scribner's Sons, New York (republished 1986, University of Wisconsin Press, Madison).

———. 1934. Conservation economics. Journal of Forestry 32:537–544.

———. 1939. A biotic view of land. Journal of Forestry 37:727–730.

———. 1941. Lakes in relation to terrestrial life patterns. Pages 17–22 in A Symposium on Hydrobiology. University of Wisconsin Press, Madison.

———. 1948. The ecological conscience. Journal of Soil and Water Conservation 3(3):109–112.

———. 1949. A Sand County almanac and sketches here and there. Oxford University Press, New York (reissued in an enlarged edition, 1966, as A Sand County almanac with essays on conservation from round river. Oxford University Press, New York; enlarged edition issued in paperback by Ballentine Books, New York).

———— 2013. Aldo Leopold: a Sand County almanac and other writings on conservation and ecology. Ed. C. Meine. Library of America 238. Library of America, New York.

Leopold, L. B., editor. 1953. Round River: from the journals of Aldo Leopold. Oxford University Press, New York.

————. 1974. Water: a primer. W.H. Freeman, San Francisco.

————. 2006. A view of the river. Harvard University Press, Cambridge, MA.

Leopold, L. B., M. G. Wolman, and J. P. Miller. 1995. Fluvial processes in geomorphology. Dover, New York.

Lewis, D. J. 2010. An economic framework for forecasting land use and ecosystem change. Resource and Energy Economics 32(2):98–116.

Lewis, D. J., and B. Provencher. 2006. The implications of heterogeneous preferences for environmental zoning. Department of Agricultural and Applied Economics, University of Wisconsin, Staff Paper No. 500, Madison.

Lewis, D. J., B. Provencher, and V. Butsic. 2009. The dynamic effects of open space conservation policies on residential development density. Journal of Environmental Economics and Management 57(3):239–252.

Lilly, L. A., B. P. Stack, and D. S. Caraco. 2012. Pollution loading from illicit sewage discharges in two Mid-Atlantic subwatersheds and implications for nutrient and bacteria total maximum daily loads. Watershed Science Bulletin 3(1):7–17.

Lindsay, A. R., S. S. Gillum, and M. W. Meyer. 2002. Influence of lakeshore development on breeding bird communities in a mixed northern forest. Biological Conservation 107:1–11.

Lodge, D. M., S. Williams, H. MacIsaac, K. Hayes, B. Leung, S. Reichard, R. N. Mack, P. B. Moyle, M. Smith, D. A. Andow, J. T. Carlton, and A. McMichael. 2006. Biological invasions: recommendations for U.S. policy and management [position paper for the Ecological Society of America]. Ecological Applications 16:2035–2054.

Ludwig, D. 1996. The end of the beginning. Ecological Applications 6:16–17.

Lyons, J. 1989. Changes in the abundance of small littoral-zone fishes in Lake Mendota, Wisconsin. Canadian Journal of Zoology 67:2910–2916.

Macbeth, E. J. 1992. Protecting aesthetics and the visual resource quality of lakes. Pages 17–23 in Enhancing the states' lake management programs. North American Lake Management Society, Washington, DC.

Mackenzie-Grieve, J. L., and J. R. Post. 2006. Thermal habitat use by lake trout in two contrasting Yukon territory lakes. Transactions of the American Fisheries Society 135:727–738.

MacRae, P. S. D., and D. A. Jackson. 2001. The influence of smallmouth bass (*Micropterus dolomieu*) predation and habitat complexity on the structure of littoral zone fish assemblages. Canadian Journal of Fisheries and Aquatic Sciences 58:342–351.

Magnuson, J. J. 2002. Signals from ice cover trends and variability. Pages 3–14 in N. A. McGinn, editor. Fisheries in a changing climate. American Fisheries Society Symposium 32, Bethesda, MD.

Magnuson, J. J., D. M. Robertson, B. J. Benson, R. H. Wynne, D. M. Livingstone, T. Arai, R. Assel, R. G. Barry, V. Card, E. Kuusisto, N. G. Granin, T. D. Prowse, K. M. Stewart, and V. S. Vuglinski. 2000. Historical trends in lake and river ice cover in the northern hemisphere. Science 289:1743–1746.

Maine Department of Environmental Protection. 1996. Lake water quality: what's it worth. Maine Department of Environmental Protection, Augusta.

Mandlebrot, B. B. 1982. The fractal geometry of nature. W.H. Freeman, New York.

Marburg, A. E., M. G. Turner, and T. K. Kratz. 2006. Natural and anthropogenic variation in coarse wood among and within lakes. Journal of Ecology 94:558–568.

Marczak, L. B., T. Sakamaki, S. L. Turvey, I. Deguise, S. L. R. Wood, and J. S. Richardson. 2010. Are forested buffers an effective conservation strategy for riparian fauna? An assessment using meta-analysis. Ecological Applications 20:126–134.

Marohn, C. L. 2012. Thoughts on building strong towns, volume 1. Charles L. Marohn, www.strongtowns.org.

Marx, L. 1964. The machine in the garden: technology and the pastoral ideal in America. Oxford University Press, NY (reissued 2000).

Mayer, P. M, S. K. Reynolds Jr., M. D. McCutchen, and T. J. Canfield. 2007. Meta-analysis of nitrogen removal in riparian buffers. Journal of Environmental Quality 36:1172–1180.

McHarg, I. 1969. Design with nature. Natural History Press, Garden City, NY (reissued 1995, Wiley, Hoboken, NJ).

McIntyre, J. W. 1988. The common loon: spirit of northern lakes. University of Minnesota Press, Minneapolis.

McKenzie-Mohr, D. 2011. Fostering sustainable behavior: an introduction to community-based social marketing, 3rd edition. New Society Publishers, Gabriola Island, BC.

McKenzie-Mohr, D, N. R. Lee, P. W. Schultz, and P. A. Kotler. 2011. Social marketing to protect the environment: what works. Sage Publications, Thousand Oaks, CA.

Mead, K. 2009. Dragonflies of the north woods. Kollath-Stensaas Publishing, Duluth, MN.

Meadows, D. H. 2007. Leverage points: places to intervene in a system. Sustainability Institute, Hartland, VT.

———. 2008. Thinking in systems: a primer. Chelsea Green Publishing, White River Junction, VT.

Meadows, D. H., D. L. Meadows, J. Randers, and W. W. Behrens III. 1972. The limits to growth. Universe Books, Boca Raton, FL. Updated in 1992 (as Beyond the limits) and 2004 (as D. H. Meadows, D. L. Meadows, and J. Randers. Limits to growth: the 30-year update. Chelsea Green Publishing, White River Junction, VT).

Meadows, G. A., S. D. Mackey, R. R. Goforth, D. M. Mickelson, T. B. Edil, J. Fuller, D. E. Guy, L. A. Meadows, E. Brown, S. M. Carmen, and D. L. Liebenthal. 2005. Cumulative habitat impacts of nearshore enginnering. Journal of Great Lakes Research 31(S1):90–112.

Meine, C. 1988. Aldo Leopold, his life and work. University of Wisconsin Press, Madison.

———. 2004. Correction lines: essays on land, Leopold, and conservation. Island Press, Washington, DC.

Meine, C., and R. L. Knight, editors. 1999. The essential Aldo Leopold. University of Wisconsin Press, Madison.

Meyer, M., J. Woodford, S. Gillum, and T. Daulton. 1997. Shoreland zoning regulations do not adequately protect wildlife habitat in northern Wisconsin. U.S. Fish and Wildlife Service, State Partnership Grant P-1-W, Segment 17, Final Report, Madison, WI.

Michael, H. J., K. J. Boyle, and R. Bouchard. 1996. Water quality affects property prices: a case study of selected Maine lakes. Maine Agricultural and Forest Experiment Station. Miscellaneous Report 398, Orono.

Milder, J. C. 2007. A framework for understanding conservation development and its ecological implications. BioScience 57:757–768.

Milder, J. C., and S. Clark. 2011. Conservation development practices, extent, and land-use effects in the United States. Conservation Biology 25(4):697–707.

Milder, J. C., J. P. Lassoie, and B. L. Bedford. 2008. Conserving biodiversity and ecosystem function through limited development: an empirical evaluation. Conservation Biology 22(1):70–79.

Milesi, C., S. W. Running, C. D. Elvidge, J. B. Dietz, B. T. Tuttle, and R. R. Nemani. 2005. Mapping and modeling the biogeochemical cycling of turf grasses in the United States. Environmental Management 36:426–438.

Miller, J. R. 2005. Biodiversity conservation and the extinction of experience. Trends in Ecology and Evolution 20:430–434.

———. 2006. Restoration, reconciliation, and reconnecting with nature nearby. Biological Conservation 127:356–361.

———. 2008. Conserving biodiversity in metropolitan areas. Landscape Journal 27:114–126.

Miller, J. R., and R. J. Hobbs. 2002. Conservation where people live. Conservation Biology 16(2):330–337.

Minnesota DNR (Department of Natural Resources). 1989. Shoreland management standards: statement of need and reasonableness. Minnesota Department of Natural Resources, St. Paul.

———. 2010. Score your shore: citizen shoreline description survey (version 2). Division of Ecological and Water Resources, Minnesota Department of Natural Resources, St. Paul.

———. 2011. Minnesota's sensitive lakeshore identification manual: a conservation strategy for Minnesota lakeshores (version 3). Division of Ecological and Water Resources, Minnesota Department of Natural Resources, St. Paul.

Minnesota PCA (Pollution Control Agency). 1999. Effects of septic systems on ground water quality—Baxter, Minnesota. Pollution Control Agency, St. Paul.

———. 2000. Ground water quality under three unsewered subdivisions in Minnesota. Pollution Control Agency, St. Paul.

———. 2001. Minnesota's nonpoint source management program plan. Pollution Control Agency, St. Paul.

———. 2004a. Minnesota lake water quality assessment data: 2004. Pollution Control Agency, St. Paul.

———. 2004b. 10-year plan to upgrade and maintain Minnesota's on-site (ISTS) treatment systems. Report to Legislature. Pollution Control Agency, St. Paul.

———. 2004c. Detailed assessment of phosphorus sources to Minnesota watersheds. Prepared by Barr Engineering. Pollution Control Agency, St. Paul.

Moilanen, A., H. Kujala, and J. Leathwick. 2009. The Zonation framework and software for conservation prioritization. Pages 196–210 in A. Moilanen, K. A. Wilson, and H. P. Possingham, editors. Spatial conservation prioritization: quantitative methods and computational tools. Oxford University Press, Oxford.

Moscovitch, E. 2007. The economic impact of proximity to open space on single-family home values in Washington County, Minnesota. Embrace Open Space, St. Paul, MN.

Moyle, J. B. 1945. Some chemical factors influencing the distribution of aquatic plants in Minnesota. American Midland Naturalist 34:402–420.

———. 1956. Relationships between the chemistry of Minnesota surface waters and wildlife management. Journal of Wildlife Management 20:303–319.

Mundell, J., S. J. Taff, M. A. Kilgore, and S. A. Snyder. 2010. Using real estate records to assess forest land parcelization and development: a Minnesota case study. Landscape and Urban Planning 94:71–76.

Murphy, K. J., and J. W. Eaton. 1983. Effects of pleasure-boat traffic on macrophyte growth in canals. Journal of Applied Ecology 20:713–729.

Nassauer, J. I., S. E. Kosek, and R. C. Corry. 2001. Public expectations with ecological innovation in riparian landscapes. Journal of the American Water Resources Association 37(6):1439–1443.

Nassauer, J. I., J. D. Allan, T. Johengen, S. E. Kosek, and D. Infante. 2004. Exurban residential subdivision development: effects on water quality and public perception. Urban Ecosystems 7:267–281.

National Park Service. 1993. Economic impacts of protecting rivers, trails, and greenway corridors: a resource book. Rivers, Trails and Conservation Assistance Section. National Park Service, Washington, DC.

National Research Council. 2009. Urban stormwater management in the United States. National Academies Press, Washington, DC.

Nelson, A., L. Bowles, J. Juergensmeyer, and J. Nicholas. 2008. A guide to impact fees and housing affordability. Island Press, Washington, DC.

Nelson, G., S. Campbell, and P. Wozniak. 2002. Beyond Earth Day: fulfilling the promise. University of Wisconsin Press, Madison.

Newbrey, J. L., M. A. Bozek, and N. D. Niemuth. 2005a. Effects of lake characteristics and human disturbance on the presence of piscivorous birds in northern Wisconsin, U.S.A. Waterbirds 28(4):478–486.

Newbrey, M. G., M. A. Bozek, M. J. Jennings, and J. E. Cook. 2005b. Branching complexity and morphological characteristics of coarse woody structure as lacustrine fish habitat. Canadian Journal of Fisheries and Aquatic Sciences 62:2110–2123.

Newman, P., T. Beatley, and H. Boyer. 2009. Resilient cities: responding to peak oil and climate change. Island Press, Washington, DC.

Nichols, S. A., and J. G. Vennie. 1991. Attributes of Wisconsin lake plants. Information Circular 73, Wisconsin Geological and Natural History Survey, Madison, WI.

Nielsen, A., D. Trolle, M. Sondergaard, T. L. Lauridsen, R. Bjerring, J. E. Olesen, and E. Jeppesen. 2012. Watershed land use effects on lake water quality in Denmark. Ecological Applications 22:1187–1200.

Nordahl, D. 2009. Public produce: the new urban agriculture. Island Press, Washington, DC.

Novotny, E. V., D. Murphy, and H. G. Stefan. 2008. Increase of urban lake salinity by road deicing salt. Science of the Total Environment 406:131–144.

O'Dell, K. M., J. VanArman, B. H. Welch, and S. D. Hill. 1995. Changes in water chemistry in a macrophyte dominated lake before and after herbicide treatment. Lake and Reservoir Management 11:311–316.

Olson, S. F. 1936. The romance of portages. Minnesota Conservationist (April). Available: http://www4.uwm.edu/letsci/research/sigurd_olson/articles/1930s/1936–04–00—The%20Romance%20 0f%20Portages—Minnesota%20Conservationist.htm (August 2013).

———. 1956. The singing wilderness. Alfred A. Knopf, New York (reissued 1997, University of Minnesota Press, Minneapolis).

———. 1958. Listening Point. Alfred A. Knopf, New York (reissued 1997, University of Minnesota Press, Minneapolis).

———. 1961. The lonely land. Alfred A. Knopf, New York (reissued 1997, University of Minnesota Press, Minneapolis).

———. 1963. Runes of the north. Alfred A. Knopf, New York (reissued 1997, University of Minnesota Press, Minneapolis).

———. 1967. Wilderness canoe country: Minnesota's greatest recreational asset. Naturalist 19 (Spring). Available: http://listeningpointfoundation.org/wilderness-canoe-country-minnesotas-greatest -recreational-asset/ (August 2013).

———. 1969a. The hidden forest. Viking Press, New York.

———. 1969b. Open horizons. Alfred A. Knopf, New York (reissued 1998, University of Minnesota Press, Minneapolis).

———. 1972. Wilderness days. Alfred A. Knopf, New York (reissued 2012, University of Minnesota Press, Minneapolis).

———. 1976. Reflections from the north country. Alfred A. Knopf, New York (reissued 1998, University of Minnesota Press, Minneapolis).

———. 1982. Of time and place. Alfred A. Knopf, New York (reissued 1998, University of Minnesota Press, Minneapolis).

Owens, D. W., P. Jopke, D. W. Hall, J. Balousek, and A. Roa. 2000. Soil erosion from two small construction sites, Dane County, Wisconsin. U.S. Geological Survey, Water Resources Investigations Report FS-109-00.

Paddock, J. 2001. Keeper of the wild: the life of Ernest Oberholtzer. Minnesota Historical Society Press, St. Paul.

Palmer, M. E., N. D. Yan, A. M. Paterson, and R. E. Girard. 2011. Water quality changes in south-central Ontario lakes and the role of local factors in regulating lake response to regional stressors. Canadian Journal of Fisheries and Aquatic Sciences 68:1038–1050.

Papenfus, M., and B. Provencher. 2006. A hedonic analysis of environmental zoning: lake classification in Vilas County, Wisconsin. Department of Agricultural and Applied Economics, University of Wisconsin, Staff Paper, Madison.

Paul, M., and J. Meyer. 2001. Streams in an urban landscape. Annual Review of Ecological Systems 32:333–365.

Payton, M. A., and D. C. Fulton. 2004. A study of landowner perceptions and opinions of aquatic plant management in Minnesota lakes. U.S. Geological Survey, Minnesota Cooperative Fish and Wildlife Research Unit. University of Minnesota, Department of Fisheries, Wildlife, and Conservation Biology, St. Paul.

Pierce, R. B. 2012. Northern pike: ecology, conservation, and management history. University of Minnesota Press, Minneapolis.

Platt, R. H., editor. 2006. The humane metropolis: people and nature in the 21st-century city. University of Massachusetts Press, Boston.

Powell, J. W. 2004. Seeing things whole: the essential John Wesley Powell. Ed. W. deBuys. Island Press, Washington, DC.

Powell, K. I., J. M. Chase, and T. M. Knight. 2013. Invasive plants have scale-dependent effects on diversity by altering species-area relationships. Science 339:316–318.

Prince, H. 1997. Wetlands of the American Midwest: a historical geography of changing attitudes. University of Chicago Press, Chicago.

Ptacek, C. 1998. Geochemistry of a septic-system plume in a coastal barrier bar, Point Pelee, Ontario, Canada. Journal of Contaminant Hydrology 33:293–312.

Pulliam, H. R., and N. M. Haddad. 1994. Human population growth and the carrying capacity concept. Bulletin of the Ecological Society of America 75(3):141–157.

Pyle, R. M. 1993. The thunder tree: lessons from an urban wildland. Houghton Mifflin, Boston.

———. 2003. Nature matrix: reconnecting people and nature. Oryx 37(2):206–214.

Radeloff, V. C., R. B. Hammer, P. R. Voss, A. E. Hagen, D. R. Field, and D. J. Mladenoff. 2001. Human demographic trends and landscape level forest management in the northwest Wisconsin pine barrens. Forest Science 47(2):229–241.

Radomski, P. 2006. Historical changes in abundance of floating-leaf and emergent vegetation in Minnesota lakes. North American Journal of Fisheries Management 26:932–940.

Radomski, P., L. A. Bergquist, M. Duval, and A. Williquett. 2010. Potential impacts of docks on littoral habitats in Minnesota lakes. Fisheries 35(10):489–495.

Radomski, P., K. Carlson, and K. Woizeschke. 2013. Common loon (*Gavia immer*) nesting habitat models for Minnesota lakes for north-central Minnesota lakes. Waterbirds (in press).

Radomski, P., and T. J. Goeman. 1995. The homogenizing of Minnesota lake fish assemblages. Fisheries 20:20–23.

———. 2001. Consequences of human lakeshore development on emergent and floating-leaf vegetation abundance. North American Journal of Fisheries Management 21:46–61.

Radomski, P., and D. Perleberg. 2012. Application of a versatile aquatic macrophyte integrity index for Minnesota lakes. Ecological Indicators 20:252–268.

Rahel, F. J. 2002a. Homogenization of freshwater faunas. Annual Review of Ecology and Systematics 33:291–315.

———. 2002b. Using current biogeographic limits to predict fish distributions following climate change. Pages 99–110 in N.A. McGinn editor. Fisheries in a changing climate. American Fisheries Society Symposium 32. Bethesda, MD.

Ramstack, J. M., S. C. Fritz, and D. R. Engstrom. 2004. Twentieth-century water quality trends in Minnesota lakes compared with presettlement variability. Canadian Journal of Fisheries and Aquatic Sciences 61:561–576.

Reed, J. R., and D. L. Pereira. 2009. Relationships between shoreline development and nest selection by black crappie and largemouth bass. North American Journal of Fisheries Management 29:943–948.

Relph, E. 1976. Place and placelessness. Pion, London.

Remsburg, A. J., and M. G. Turner. 2009. Aquatic and terrestrial drivers of dragonfly (*Odonata*) assemblages within and among north-temperate lakes. Journal of the North American Benthological Society 28(1):44–56.

Rittel, H. W., and M. M. Webber. 1973. Dilemmas in a general theory of planning. Policy Sciences 4:155–169.

Robertson, R. J., and N. J. Flood. 1980. Effects of recreational use of shorelines on breeding bird populations. Canadian Field-Naturalist 94:131–138.

Robertson, W. D. 1995. Development of steady-state phosphate concentrations in septic system plumes. Journal of Contaminant Hydrology 19:289–305.

———. 2003. Enhanced attenuation of septic system phosphate in noncalcareous sediments. Ground Water 41(1):48–56.

———. 2008. Irreversible phosphorus sorption in septic system plumes? Ground Water 46(1):51–60.

Robertson, W. D., S. Schiff, and C. Ptaceck. 1998. Review of phosphate mobility and persistence in 10 septic system plumes. Ground Water 36(6):1000–1010.

Rogers, E. M. 2003. Diffusion of innovations, 5th edition. Free Press, New York.

Rosenzweig, M. L. 2003. Win-win ecology: how the Earth's species can survive in the midst of human enterprise. Oxford University Press, New York.

Roth, B. M., I. C. Kaplan, G. G. Sass, P. T. Johnson, A. E. Marburg, A. C. Yannarell, T. D. Havlicek, T. V. Willis, M. G. Turner, and S. R. Carpenter. 2007. Linking terrestrial and aquatic ecosystems: the role of woody habitat in lake food webs. Ecological Modelling 203:439–452.

Ruhl, J. B., S. E. Kraft, and C. L. Lant. 2007. The law and policy of ecosystem services. Island Press, Washington, DC.

Ruhl, J. B., C. Lant, T. Loftus, S. Kraft, J. Adams, and L. Duram. 2003. Proposal for a model state watershed management act. Environmental Law 33:929–947.

Rutstrum, C. 1958. The new way of the wilderness: the classic guide to survival in the wild. Macmillan, New York (republished 2000, University of Minnesota Press, Minneapolis).

Sandel, M. J. 2012. What money can't buy: the moral limits of markets. Farrar, Straus, and Giroux, New York.

Sander, H. A., and S. Polasky. 2009. The value of views and open space: estimates from a hedonic pricing model for Ramsey County, Minnesota, USA. Land Use Policy 26(3):837–845.

Sanders, S. R. 1993. Staying put: making a home in a restless world. Beacon Press, Boston.

Sanzo, D., and S. J. Hecnar. 2006. Effects of road de-icing salt (NaCl) on larval wood frogs (*Rana sylvatica*). Environmental Pollution 140:247–256.

Sax, J. L. 1970. The public trust doctrine in natural resource law: effective judicial intervention. Michigan Law Review 68(3):471–566.

———. 1989. The limits of private rights in public waters. Environmental Law 19:473–483.

———. 1996. Using property rights to attack environmental protection: second annual Lloyd K. Garrison lecture on environmental law. Pace Environmental Law Review 14(1):1–14.

———. 2005. Why America has a property rights movement. University of Illinois Law Review 2005(2):513–520.

———. 2010. Land use regulation: time to think about fairness. Natural Resources Journal 50:455–470.

———. 2011. Ownership, property, and sustainability. Utah Environmental Law Review 31(1):1–16.

Scheffer, M., and S. R. Carpenter. 2003. Catastrophic regime shifts in ecosystems: linking theory to observation. Trends in Ecology and Evolution 18:648–656.

Scheffer, M., S. H. Hosper, M.-L. Meijer, B. Moss, and E. Jeppesen. 1993. Alternative equilibria in shallow lakes. Trends in Ecology and Evolution 8:275–279.

Scheffer, M., and E. H. van Nes. 2007. Shallow lakes theory revisited: various alternative regimes driven by climate, nutrients, depth and lake size. Hydrobiologia 584:455–466.

Scheuhammer, A. M. 2009. Historical perspective on the hazards of environmental lead from ammunition and fishing weights in Canada. Pages 61–67 in R. T. Watson, M. Fuller, M. Pokras, and W. G. Hunt, editors. Ingestion of lead from spent ammunition: implications for wildlife and humans. Peregrine Fund, Boise, ID.

Schindler, D. W. 2001. The cumulative effects of climate warming and other human stresses on Canadian freshwaters in the new millennium. Canadian Journal of Fisheries and Aquatic Sciences 58:18–29.

———. 2006. Recent advances in the understanding and management of eutrophication. Limnology and Oceanography 51:56–363.

Schindler, D. W., R. E. Hecky, D. L. Findlay, M. P. Stainton, B. R. Parker, M. J. Paterson, K. G. Beaty, M. Lyng, and S. E. M. Kasian. 2008. Eutrophication of lakes cannot be controlled by reducing nitrogen input: results of a 37-year whole-ecosytem experiment. Proceedings of the National Academy of Sciences of the United States of America 105:11254–11258.

Schindler, D. E., and M. D. Scheuerell. 2002. Habitat coupling in lake ecosystems. Oikos 98:177–189.

Schlaepfer, M. A., D. F. Sax, and J. D. Olden. 2011. The potential conservation value of non-native species. Conservation Biology 25(3):428–437.

———. 2012. Toward a more balanced view of non-native species. Conservation Biology 26(6):1156–1158.

Schoenbauer, J. 2007. Redefining the development process handbook. Brauer & Accociates and Schoenbauer Consulting, Minneapolis, MN.

Schroeder, S. A. 2009. Quality connections: recreation, property ownership, place attachment, and conservation of Minnesota lakes. Doctoral dissertation. University of Minnesota, Department of Fisheries, Wildlife, and Conservation Biology, St. Paul.

Schroeder, S., M. A. Payton, and D. C. Fulton. 2004. A study of the general public's perceptions and opinions of lake and aquatic management in Minnesota. U.S. Geological Survey, Minnesota Cooperative Fish and Wildlife Research Unit. University of Minnesota, Department of Fisheries, Wildlife, and Conservation Biology, St. Paul.

Schroeder, S. A., and D. C. Fulton. 2013. Public lakes, private lakeshore: modeling protection of native aquatic plants. Environmental Management 52:99–112.

Schueler, T. 1994. The importance of imperviousness. Watershed Protection Techniques 1(3): 100–111.

———. 2000. An introduction to better site design. Watershed Protection Techniques 3(2):623–632.

———. 2003. Impacts of impervious cover on aquatic systems. Watershed Protection Research Monograph 1. Center for Watershed Protection, Ellicott City, MD.

———. 2008. Technical support for the bay-wide runoff reduction method. Chesapeake Stormwater Network, Baltimore, MD.

Schueler, T. R., and D. S. Caraco. 2001. The prospects for low impact land development at the watershed level. Pages 196–209 in Linking stormwater BMP designs and performance to receiving water impacts mitigation. United Engineering Foundation, Snowmass, CO.

Schueler, T. R., L. Fraley-McNeal, and K. Cappiella. 2009. Is impervious cover still important? Review of recent research. Journal of Hydrologic Engineering 14(4):309–315.

Schueler, T. R., and H. K. Holland. 2000. The practice of watershed protection. Center for Watershed Protection, Ellicott City, MD.

Schussler, J., L. A. Baker, and H. Chester-Jones. 2007. Whole-system phosphorus balances as a practical tool for lake management. Ecological Engineering 29:294–304.

Selbig, W. R., and N. Balster. 2010. Evaluation of turf-grass and prairie-vegetated rain gardens in a clay and sand soil: Madison, Wisconsin, water years 2004–08. U.S. Geological Survey, Scientific Investigations Report 2010-5077.

Selbig, W. R., and R. T. Bannerman. 2008. A comparison of runoff quantity and quality from two small basins undergoing implementation of conventional- and low-impact-development (LID) strategies: Cross Plains, Wisconsin, water years 1999–2005. U.S. Geological Survey, Scientific Investigations Report 2008-5008.

Senge, P. M., A. Kleiner, C. Roberts, R. B. Ross, and B. J. Smith. 1994. The fifth discipline fieldbook: strategies and tools for building a learning organization. Double Day, New York.

Sharma, S., P. Legendre, D. Boisclair, and S. Gauthier. 2012. Effects of spatial scale and choice of statistical model (linear versus tree-based) on determining species-habitat relationships. Canadian Journal of Fisheries and Aquatic Sciences 69:2095–2111.

Skawinski, P. M. 2011. Aquatic plants of the upper Midwest: a photographic field guide to our underwater forests. Paul M. Skawinski, Stevens Point, WI.

Skinner, B. F. 1987. Upon further reflection. Prentice-Hall, Englewood Cliffs, NJ.

Smith, M., D. de Groot, and G. Bergkamp. 2006. Pay: establishing payments for watershed services. International Union for Conservation of Nature and Natural Resources, Gland, Switzerland.

Snyder, G. 1990. The practice of the wild: Essays. North Point Press, New York.

Soranno, P. A., K. S. Cheruvelil, R. J. Stevenson, S. L. Rollins, S. W. Holden, S. Heaton, and E. Torng. 2008. A framework for developing ecosystem-specific nutrient criteria: integrating biological thresholds with predictive modeling. Limnology and Oceanography 53:773–787.

Soranno, P. A., K. S. Cheruvelil, K. E. Webster, M. T. Bremigan, T. Wagner, and C. A. Stow. 2010.

Using landscape limnology to classify freshwater ecosystems for multi-ecosystem management and conservation. BioScience 60:440–454.

Spalatro, F., and B. Provencher. 2001. Analysis of minimum frontage zoning to preserve lakefront amenities. Land Economics 77(4):469–481.

Speck, J. 2012. Walkable city: how downtown can save America, one step at a time. Farrar, Straus and Giroux, New York.

Stedman, R. C. 2003. Is it really just a social construction: the contribution of the physical environment to sense of place. Society and Natural Resources 16:671–685.

———. 2006. Understanding place attachment among second home owners. American Behavioral Scientist 50(2):187–205.

Stedman, R. C., and R. B. Hammer. 2006. Environmental perception in a rapidly growing, amenity-rich region: the effects of lakeshore development on perceived water quality in Vilas County, Wisconsin. Society and Natural Resources 19:137–151.

Stedman, R. C., R. C. Lathrop, B. Clark, J. Ejsmont-Karabin, P. Casprzak, K. Nielsen, D. Osgood, M. Powell, A. Ventelä, K. E. Webster, and A. Zhukova. 2007. Perceived environmental quality and place attachment in North American and European temperate lake districts. Lake and Reservoir Management 23:330–344.

Stefan, H. G., M. Hondzo, X. Fang, J. G. Eaton, and J. H. McCormick. 1996. Simulated long-term temperature and dissolved oxygen characteristics of lakes in the north-central United States and associated fish habitat limits. Limnology and Oceanography 41:1124–1135.

Stefanovic, I. L. 2012. To build or not to build a road: how do we honor the landscape through thoughtful decision making? Minding Nature 5(1):12–18.

Steffy, L. Y., and S. S. Kilham. 2004. Elevated δ15N in stream biota in areas with septic tank systems in an urban watershed. Ecological Applications 14:637–641.

Stegner, W. 1992. Where the bluebird sings to the lemonade springs: living and writing in the west. Penquin, New York.

Stone, B. 2004. Paving over paradise: how land use regulations promote residential imperviousness. Landscape and Urban Planning 69:101–113.

Strayer, D. L. 2010. Alien species in fresh waters: ecological effects, interactions with other stressors, and prospects for the future. Freshwater Biology 55(Suppl. 1):152–174.

Swift, P., D. Painter, and M. Goldstein. 2006. Residential street typology and injury accident frequencies. Swift and Associates, Longmont, CO.

Szold, T. S., and A. Carbonell, editors. 2002. Smart growth: form and consequences. Lincoln Institute of Land Policy, Cambridge, MA.

Tabor, R. A., K. L. Fresh, R. M. Piaskowski, H. A. Gearns, and D. B. Hayes. 2011. Habitat use by juvenile Chinook salmon in the nearshore areas of Lake Washington: effects of depth, lakeshore development, substrate, and vegetation. North American Journal of Fisheries Management 31:700–713.

Tainter, J. A. 1990. The collapse of complex societies, 1st paperback edition. Cambridge University Press, Cambridge.

Thaler, R. H., and C. R. Sunstein. 2008. Nudge: improving decisions about health, wealth, and happiness. Yale University Press, New Haven, CT.

Thompson, J. W., and K. Sorvig. 2008. Sustainable landscape construction: a guide to green building outdoors. Island Press, Washington, DC.

Titus, J. R., and L. W. VanDruff. 1981. Response of the common loon to recreational pressure in the Boundary Waters Canoe Area, northeastern Minnesota. Wildlife Monographs 79. Wildlife Society, Washington, DC.

Trial, P. F., F. P. Gelwick, and M. A. Webb. 2001. Effects of shoreline urbanization on littoral fish assemblages. Lake and Reservoir Management 17:127–138.

Trust for Public Land. 2009. Hennepin County economic analysis. Trust for Public Land, St. Paul, MN.

Tuan, Y.-F. 1977. Space and place: the perspective of experience. University of Minnesota Press, Minneapolis.

U.S. Census Bureau. 2005. 2004 population estimates for micropolitan and metropolitan statistical areas. U.S. Census Bureau, Washington, DC. Available: http://www.census.gov/popest/data/historical/2000s/vintage_2005/metro.html (August 2013)

U.S. EPA (Environmental Protection Agency). 1994. National water quality inventory. 1992 Report to Congress. Publication EPA-841-R-94-001, Washington, DC.

———. 1998. The quality of our nation's waters: a summary of the national water quality inventory. 1998 Report to Congress. Publication EPA-841-S-00-001, Washington, DC.

———. 2000. Low impact development: a literature review. Publication EPA-841-B-00-005, Washington, DC.

———. 2009. National lakes assessment: a collaborative survey of the nation's lakes. Publication EPA-841-R-09-001, Office of Water and Office of Research and Development, Washington, DC.

———. 2010. Green infrastructure case studies: municipal policies for managing stormwater with green infrastructure. Publication EPA-841-F-10-004, Office of Wetlands, Oceans, and Watersheds, Washington, DC.

———. 2012. Evaluation of combined heat and power technologies for wastewater facilities. Publication EPA-832-R-10-006, Washington, DC.

U.S. EPA and Combined Heat and Power Partnership. 2011. Opportunities of combined heat and power at wastewater treatment facilities: market analysis and lessons from the field. U.S. Environmental Protection Agency, Washington, DC. Available: http://www.epa.gov/chp/documents/wwtf_opportunities.pdf (August 2013).

Valley, R. D., T. K. Cross, and P. Radomski. 2004. The role of submersed aquatic vegetation as habitat for fish in Minnesota lakes, including the implications of non-native plant invasions and their management. Minnesota Department of Natural Resources, Special Publication 160, St. Paul.

Valley, R. D., W. Crowell, C. H. Welling, and N. Proulx. 2006. Effects of a low dose fluridone treatment on submersed aquatic vegetation in a eutrophic Minnesota lake dominated by Eurasian watermilfoil and coontail. Journal of Aquatic Plant Management 44:19–25.

Valley, R. D., M. D. Habrat, E. D. Dibble, and M. T. Drake. 2010. Movement patterns and habitat use of three declining littoral fish species in a north-temperate mesotrophic lake. Hydrobiologia 644:385–399.

Van Assche, K. 2013. Ernest Oberholtzer and the art of boundary crossing: writing, life and the narratives of conservation and planning. Planning Perspectives (in press).

Van Assche, K., R. Beunen, M. Duineveld, and H. De Jong. 2013. Co-evolutions of planning and design: risks and benefits of design perspectives in planning systems. Planning Theory 12:177–198.

Van Assche, K., M. Duineveld, H. De Jong, and A. Van Zoest. 2012. What place is this time? Semiotics and the analysis of historical reference in landscape architecture. Journal of Urban Design 17:233–254.

Van Assche, K., and K. Verschraegen. 2008. The limits of planning: Niklas Luhmann's systems theory and the analysis of planning and planning ambitions. Planning Theory 7:263–283.

Van Hise, C. 1910. The conservation of natural resources in the United States. Macmillan, New York.

Vander Zanden, M. J., and C. Gratton. 2011. Blowin' in the wind: reciprocal airborne carbon fluxes between lakes and land. Canadian Journal of Fisheries and Aquatic Sciences 68:170–182.

Vander Zanden, M. J., J. D. Olden, J. H. Thorne, and N. E. Mandrak. 2004. Predicting the occurrence and impact of bass introductions in north-temperate lakes. Ecological Applications 14:132–148.

Vermeer, K. 1973. Some aspects of the breeding and mortality of common loons in east-central Alberta. Canadian Field-Naturalist 87:403–408.

Vitousek, P. M., P. R. Ehrlich, A. H. Ehrlich, and P. A. Matson. 1986. Human appropriation of the products of photosynthesis. BioScience 36:368–373.

Vitousek, P. M., H. A. Mooney, J. Lubchenco, and J. M. Melillo. 1997. Human domination of Earth's ecosystems. Science 227:494–499.

Wagner, K. I., J. Hauxwell, P. W. Rasmussen, F. Koshere, P. Toshner, K. Aron, D. R. Helsel, S. Toshner, S. Provost, M. Gansberg, J. Masterson, and S. Warwick. 2007. Whole-lake herbicide treatments for Eurasian watermilfoil in four Wisconsin lakes: effects on vegetation and water clarity. Lake and Reservoir Management 23:83–94.

Wang, L., and P. Kanehl. 2003. Influences of watershed urbanization and instream habitat on macroinvertebrates in cold water streams. Journal of the American Water Resources Association 39(5):1181–1196.

Wang, L., J. Lyons, and P. Kanehl. 2001. Impacts of urbanization on stream habitat and fish across multiple spatial scales. Environmental Management 28:255–266.

———. 2003. Impacts of urban land cover on trout streams in Wisconsin and Minnesota. Transactions of the American Fisheries Society 132:825–839.

Wang, L., J. Lyons, P. Kanehl, R. Bannerman, and E. Emmons. 2000. Watershed urbanization and changes in fish communities in southeastern Wisconsin streams. Journal of the American Water Resources Association 36(5):1173–1189.

Wang, L., J. Lyons, P. Kanehl, and R. Gatti. 1997. Influences of watershed land use on habitat quality and biotic integrity in Wisconsin streams. Fisheries 22:6–12.

Waschbusch, R. J., W. R. Selbig, and R. T. Bannerman. 1999. Sources of phosphorus in stormwater and street dirt from two urban residential basins in Madison, Wisconsin, 1994–95. U.S. Geological Survey, Water Resources Investigations Report 99-4021.

Wehrly, K. E., J. E. Breck, L. Wang, and L. Szabo-Kraft. 2012. Assessing local and landscape patterns of residential shoreline development in Michigan lakes. Lake and Reservoir Management 28:158–169.

Wei, A., P. Chow-Fraser, and D. Albert. 2004. Influence of shoreline features on fish distribution in the Laurentian Great Lakes. Canadian Journal of Fisheries and Aquatic Sciences 61:1113–1123.

Weigel, B. M., E. E. Emmons, J. S. Stewart, and R. Bannerman. 2005. Buffer width and continuity for preserving stream health in agricultural landscapes. Research/Management Findings 56, Wisconsin Department of Natural Resources, Madison.

Wenger, S. 1999. A review of the scientific literature on riparian buffer width, extent and vegetation. Office of Public Services and Outreach, Institute of Ecology, University of Georgia, Athens.

Wenger, S., and L. Fowler. 2000. Protecting stream and river corridors: creating effective local riparian buffer ordinances. Carl Vinson Institute of Government, University of Georgia, Athens.

———. 2001. Conservation subdivision ordinances. Office of Public Service and Outreach for the Atlanta Regional Commission, Institute of Ecology, University of Georgia, Athens.

Wenger, S. J., J. T. Peterson, M. C. Freeman, B. J. Freeman, and D. D. Homans. 2008. Stream fish occurrence in response to impervious surface cover, historic land use, and hydrogeomorphic factors. Canadian Journal of Fisheries and Aquatic Sciences 65:1250–1264.

Wetzel, R. G. 2001. Limnology: lake and river ecosystems, 3rd edition. Academic Press, San Diego, CA.

Whyte, W. H., Jr. 1956. The organization man. Simon and Schuster, New York (republished 2002, University of Pennsylvania Press, Philadelphia).

———. editor. 1958. The exploding metropolis. Doubleday Anchor Books, Garden City, NY. (republished 1993, University of California Press, Berkeley).

———. 1959. A plan to save vanishing U.S. countryside. Life 47(7):88–102.

———. 1959. Securing open space for urban America: conservation easements. Urban Land Institute, Washington, DC.

———. 1962. Open space action: a report to the Outdoor Recreation Resources Review Commission. Study Report 15. U.S. Government Printing Office, Washington, DC.

———. 1964. Cluster development. American Conservation Association, New York.

———. 1968. The last landscape. Doubleday, New York (republished 2002, University of Pennsylvania Press, Philadelphia).

———. 1980. The social life of small urban spaces. Conservation Foundation, Washington, DC.

———. 1988. City: rediscovering the center. Doubleday, New York (republished 2002, University of Pennsylvania Press, Philadelphia).

———. 2000. A time of war: remembering Guadalcanal, a battle without maps. Fordham University Press, New York.

Wildlife Habitat Enhancement Council. 1992. The economic benefits of wildlife enhancement on corporate lands. Wildlife Habitat Enhancement Council, Silver Spring, MD.

Winks, R. W. 1997. Laurance S. Rockefeller: catalyst for conservation. Island Press, Washington, DC.

Woodard, S. E., and C. A. Rock. 1995. Control of residential stormwater by natural buffer strips. Lake and Reservoir Management 11:37–45.

Woodford, J. E., and M. W. Meyer. 2003. Impact of lakeshore development on green frog abundance. Biological Conservation 110:277–284.

World Commission on Environment and Development. 1987. Our common future. Oxford University Press, New York.

Wright, K. R. 2004. Machu Picchu: prehistoric construction and water handling. Pages 133–139 in J. R. Rogers, G. O. Brown, and J. D. Garbrecht, editors. Water resources and environmental history. American Society of Civil Engineers, Reston, VA.

Wright, K. R., A. V. Zegarra, R. M. Wright, G. McEwan. 2000. Machu Picchu: a civil engineering marvel. ASCE Press, Reston, VA.

Yousef, Y. A., W. M. McLellon, and H. H. Zebuth. 1980. Changes in phosphorus concentrations due to mixing by motorboats in shallow lakes. Water Research 14:841–852.

Zhang, C., and K. J. Boyle. 2010. The effect of an aquatic invasive species (Eurasian watermilfoil) on lakefront property values. Ecological Economics 70(2):394–404.

Zipperer, W. C., J. Wu, R. V. Pouyat, and S. T. A. Pickett. 2000. The application of ecological principles to urban and urbanizing landscapes. Ecological Applications 10:685–688.

Index